有本有源——售楼处设计

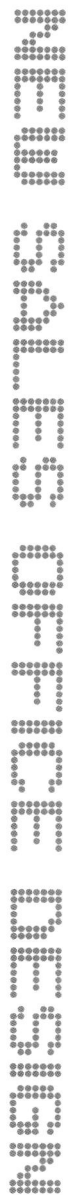

THE OPEN OPTION ROOMS

深圳市艺力文化发展有限公司 编　　大连理工大学出版社

Artpower International Publishing CO., LTD　　Dalian University of Technology Press

图书在版编目(CIP)数据

有本有源：售楼处设计：汉英对照 / 深圳市艺力
文化发展有限公司编. — 大连：大连理工大学出版社，
2011.6

ISBN 978-7-5611-6228-6

Ⅰ.①有… Ⅱ.①深… Ⅲ.①商业建筑—建筑设计—
汉、英 Ⅳ.①TU247

中国版本图书馆CIP数据核字（2011）第086968号

出版发行：大连理工大学出版社
　　　　　（地址：大连市软件园路80号　邮编：116023）
印　　　刷：利丰雅高印刷（深圳）有限公司
幅面尺寸：245mm×290mm
印　　张：21.5
插　　页：4
出版时间：2011年6月第1版
印刷时间：2011年6月第1次印刷
策　　划：袁　斌
责任编辑：刘　蓉　裘美倩　初　蕾
责任校对：张　泓　李　楠
封面设计：王　佳

ISBN 978-7-5611-6228-6
定　　价：320.00元

电　话：0411-84708842
传　真：0411-84701466
邮　购：0411-84703636
E-mail：designbooks_dutp@yahoo.cn
URL：http://www.dutp.cn

如有质量问题请联系出版中心：（0411）84709246　84709043

序言 Preface

美学经济在繁殖

上海大匀国际空间设计
协同主持人/创意总监 林宪政

> 美学的生活，就是把自己的身体、行为、感觉和激情，把自己不折不扣的存在，都变成一种艺术品。
>
> —— 法国哲学家 Michel Foucault

售楼处文化，从最单一的一张桌子，一张椅子，渐渐地变成了如今花枝招展、百花齐放的面貌，回溯到客观的商业与设计角度而言，不难看出来，售楼处文化也正代表着某种进化论。

真正的售楼处文化早已经无从考证，然而，我愿意用几个现象（或阶段）来看待这样的规律和逻辑。

1851年的伦敦世界博览会"水晶宫"的出现像是一座横空出世、前所未有的"交流馆"。当时容纳了10万多件珍宝，但却是在展期甚至于之后，水晶宫本身却成为了最重要的头号展览品。这似乎与现今的售楼处空间有着不谋而合的同质属性。

从单纯的"室内机能满足"扩展到"表皮的充分表演"，而这些历程，也代表了各个不同的时空背景以及社会价值的判断变化。

在现今各行各业都打着"美学"旗帜之际，机能美学、营销美学、创意美学、产品美学……从无到有、从抽象到具体，似乎所有的产业，都在响应"美学经济"，疯狂并铺天盖地地席卷着。对一个专业的空间设计人来说，都市发展的现象孕育着各种层面的轨迹，正如1913年，构成主义大师塔林所说"艺术家是生活风格的规划者"。我们正处于这个"困苦焦虑的风格时代"，无论是千姿百态，还是清雅飘逸。售楼处文化一时之间仍是空间前沿的探索角色，更是一种实验、验证、思考、实践、矛盾冲突、商业、摧毁的空间辩证。

售楼处的建筑哲学，可能是一种当今结合了理性、商业、潮流、时尚的复合空间综合体，但却又是最具时代代表性的微建筑。

售楼处文化，是一个城市美学的微起点，无论是透明凄美的玻璃空间，或是坚韧无比的堡垒形体，恐怕当我们30年、50年后回首，就会发现这个混杂风格代表的是一种典型的"纯粹的混合风格"。

Economy of Aesthetics is booming

Lin Xianzheng
Collaboration Moderator/Creative Director of Shanghai Symmetry International
Space Design

Aesthetic life is to turn one's own body, behavior, feeling, passion and veritable existence into an artwork.

Michel Foucault, French philosopher

Now, the real estate sales office culture has gradually turned into gorgeous and colorful looks from a single table and chair. When dating back to the objective perspective of commerce and design, it will be obvious that the real estate sales office culture also represents some kind of evolutionism.

There's no evidence to certify the genuine real estate sales office culture. However, I would like to look upon this rule and logic with several phenomena (or stages).

The emergence of 'Crystal Palace' in London World Expo 1851 seems to be a distinguished and unprecedented 'Communication Museum'. It contained more than 100,000 jewels at that time. Yet it happened during the exhibition period or afterwards that Crystal Palace itself became the primary No.1 exhibit. Its property seems to be homogeneous with that of real estate sales office space of today.

With its expansion from simple 'satisfaction of indoor function' to 'well performance of epidermis', these processes also embody diverse space-time background and variable judgment of social value.

Nowadays, all the industries are responding to the banner of 'aesthetics'. Function aesthetics, sales aesthetics, creative aesthetics and product aesthetics...from inexistence to existence, from abstraction to concreteness, all of the industries seem to respond to 'Economy of Aesthetics'. This crazily prevails in the world. However, for a professional space designer, the phenomenon of urban development has nourished the traces of diverse fields, just as what the constructivism master Tatlin Vladimirn said in 1913'artists are the planners of life styles'. Yet we are undergoing this 'era of miserable and stressful style' regardless of diversity or elegance. The real estate sales office culture still plays an important role in exploring the space frontier. Besides, it is a kind of space dialectic for experiment, verification, contemplation, practice, conflict, commerce and damage.

The building philosophy of real estate sales office may be a kind of compound-space complex combining with rationality, commerce, trend and fashion. Yet it is of micro-architecture with the typicality of an era.

The culture of real estate sales office is a new starting point of urban aesthetics. Regardless of transparent, desolate but beautiful glass space or tough fortress structure, this mixed style will represent a typical "pure mixed style" when we look back on it 30 or 50 years later.

目录 Contents

超级机器设计工作室

Pitupong Chaowakul

Pitupong Chaowakul, born in October, 1975 in Ubonratchathani, Thailand, firstly graduated from Faculty of Architecture, Chulalongkorn University, Bangkok, Thailand, where he got his Bachelor's Degree in Architecture. Then in 2002, he got his Master's Degree in The Berlage Institute of Architecture, Rotterdam, the Netherlands. He has worked in both Singapore and the Netherlands as an Architect and co-found the studio ThisDesign. In 2009, he found his own architect office Supermachine Studio.

Pitupong Chaowakul，1975年10月出生于泰国乌汶。毕业于泰国曼谷的朱拉隆功大学，取得建筑学士学位。2002年，他取得荷兰鹿特丹港市贝拉含建筑研究所的硕士学位。他以建筑师的身份工作于新加坡和荷兰，并与人合作建立工作室 ThisDesign。2009年建立自己的超级机器设计工作室。

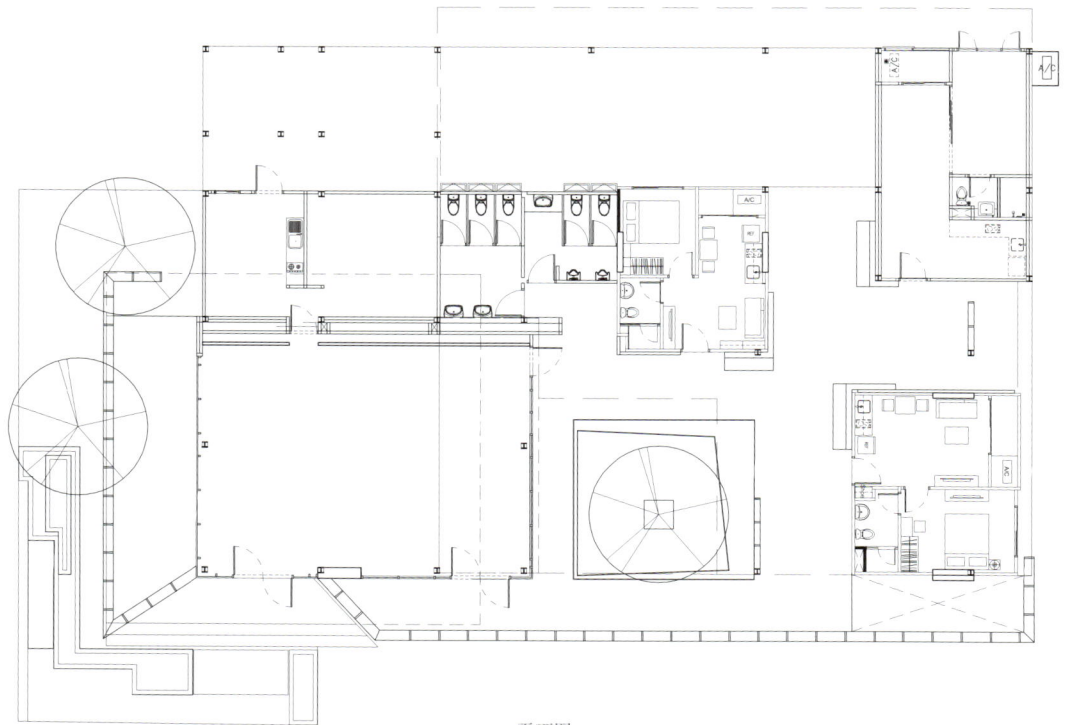

平面图

The Base Sales Gallery

Design Agency: Supermachine Studio
Location: Bangkok, Thailand
Client: Sansiri Public Company Limited
Area: 425m²
Photography: Pitupong Chaowakul

基地销售中心

设计公司：超级机器设计工作室
项目地点：泰国曼谷
客　　户：Sansiri Public Company Limited
项目面积：425m²
摄 影 师：Pitupong Chaowakul

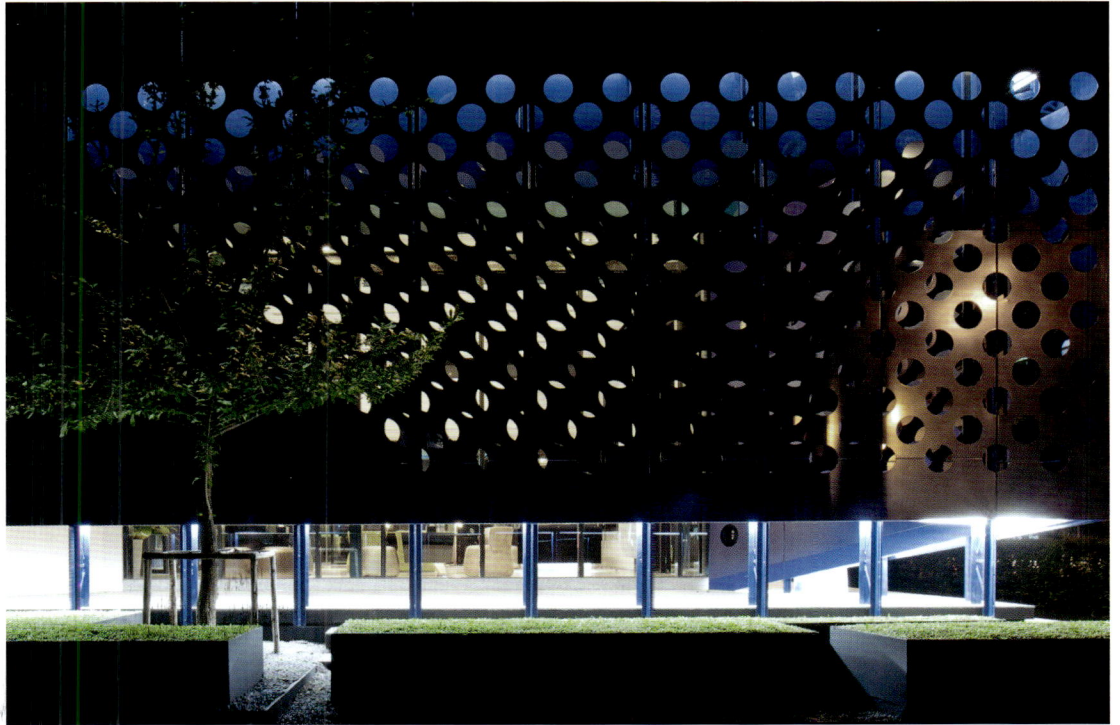

It is Supermachine's long interest in using moire pattern in architecture as the simple system that can add complexity to the building. When we have a chance, we implement it on our projects. Sansiri development came up with a small project to facelift one of their sales galleries for their new condominium in Bangkok, The Base, situating near to Ornnuch sky train station.

We were testing on several options of building skin patterns as it is to be giving different expression to the neighborhood as well as experiences to the customers. The option we choose is to wrap the building with double layer skin perforated with slightly different polka dot patterns, which we call it 'Moire Mask'. When customers or passers-by move along the building, its opening effect is changing along. The depth of the building also alters due to the changing of light throughout the day from flat skin during natural light period to a more ambivalent object when it gets darker in the day and interior illuminated.

Partly wrapping inside the courtyard, the dark blue 'Moire Mask' also define new character of the show-flat compound from a straight forward 'building around courtyard' to a more ambiguous space. Hundreds of openings on the wall with their illusion add another layer of experience onto the customers while they are wandering around the building looking at the products.

超级机器工作室一直乐于在简单的建筑中运用云绸纹来增加建筑物的复杂性。我们将此应用到我们每个项目中。此案是Sansiri公司曼谷新地产项目的一个售楼处，位于Ornnuch高架火车站附近。

为了使本建筑从周围各种建筑物中脱颖而出，同时给来此的客户一种全新的体验，我们尝试了许多种建筑外观。最终决定建筑外墙为双层有着交错波尔卡圆点的形式，我们称之为"云纹面具"。每位来访者和过路人经过时，透过外墙的光线也随之而变。建筑物的视觉深度也随着日间光线的改变而变化。在白天的自然光线下，看起来是平平的外墙；当天色逐渐暗淡，室内照明开启后，光线从室内传出，整个外墙散发着矛盾的美感。

深蓝色的"云纹面具"部分延伸至庭院，造就了一个更加模糊了界限的空间，而不是直截了当的围绕着庭院建立的建筑，赋予了售楼处新的特色。当客户漫步其中时，墙上许许多多的圆孔产生的虚幻交错之感，给他们带来另一种体验。

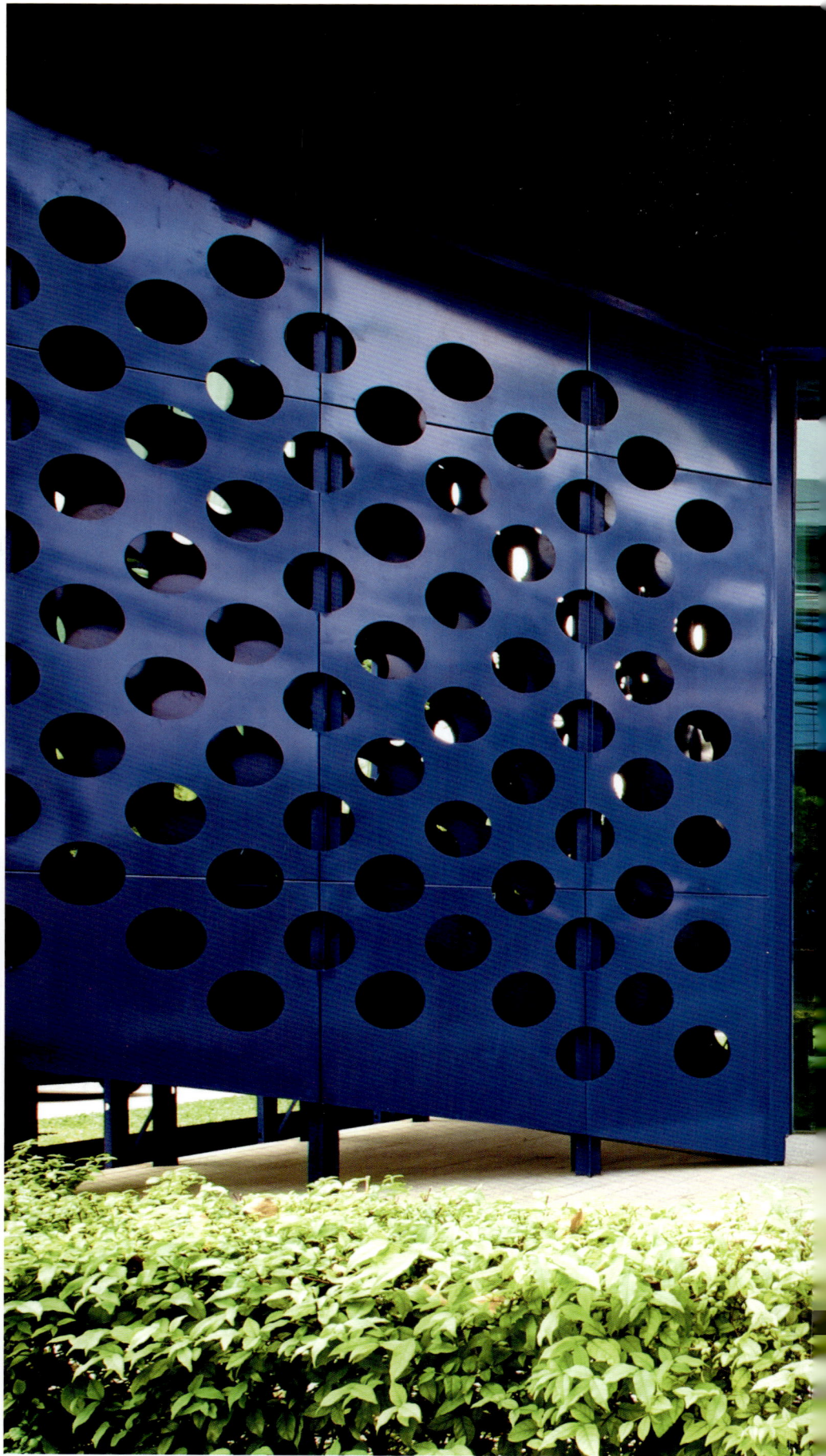

OPEN 10 AM - 6 PM EVERYDAY

Wyne Sales Gallery

Design Agency: Supermachine studio
Location: Bangkok, Thailand
Client: Sansiri Public Company Limited
Area: 720m²
Photography: Wison Tungthunya

威内销售中心

设计公司：超级机器设计工作室
项目地点：泰国曼谷
客　　户：Sansiri Public Company Limited
面　　积：720m²
摄 影 师：Wison Tungthunya

Bangkok has been witnessing one of the biggest condominium boom in its history for the past 4 years because of the success of its mass transit expansion. Along the sky-train line, architectural typologies are changing rapidly according to rise of the land price. Sansiri is one of the top land developers who is constantly re-define themselves and their products.

Wyne is one of Sansiri's latest product, a 31-storey high building in deep red color targeting to attract more than 400 younger generation families who live their lives with attitude and passion. Wyne started up passionately by redefining its sales office. Adding on to the sample of show flats, the customers can experience a subtle kind of lifestyle visiting the sales office. It is like coming to a living gallery in a park surrounded by urban elements.

The typical reception hall+office+mock up room is compressed into one chocolate bar-like box measured 30 meters long and 10 meters wide. We would like create a very simple yet challenging piece of architecture by lifting the long span box up just high enough that people can access from the space below. The supporting structure are placed in the middle accommodating the main staircase which is the main entrance to the reception hall above. This makes the building spanning 12 meters in both sides. The balancing structure cladded in brown aluminum composite panel is almost like an alien spaceship just touching the earth with its ladder. It has a transparent cylinder tube above of the entrance bringing in the natural light from the sky signifying main access as well as visually connecting the terrace and the gallery level.

Most of the windows are tangram-like geometries. This is to emphasize the characteristic of the main building in which the main highlight is the sharp age triangular "panoramic view" living room situated right in front of the building. Part of this room is also built with steel structure cantilevering out from the gallery to the back garden. The customer can have a glimpse of the visual panorama with the exact type of window frames as what they are going to buy.

因为城市轨道的成功扩张，在过去4年中曼谷见证了历史上最大的公寓热潮之一。沿着高架火车线一线，土地价格上涨，建筑类型也在迅速变化着。Sansiri是不断地重新界定自己和自己的产品的顶级土地开发商之一。

威内是Sansiri的最新产品之一，它是一栋31层的深红色建筑，专门针对400多户年轻一代的家庭，提倡富有激情的生活态度。 威内重新定义并积极地启动了销售中心。通过样品房的展示，客户可以在售楼处体会到一种微妙的生活方式，就像来到了一个城市中的公园。

售楼处30米长，10米宽，像一个被压缩成巧克力条大小的的盒子，拥有接待大厅、办公室和展示处。建筑被提升起来，人们可以在下方自由穿行，这是一个简单但具有挑战性的举动。支撑结构被设置在正中央，与主楼梯是一体的，它是去接待大厅中层的主入口， 整个建筑的下方挑空跨越了12米。外面的棕色铝复合板让建筑看上去像一个刚刚到达地球放下阶梯的外星飞船。在入口上方有一个透明的圆柱体管子，从天空中引入自然光，既标明了这里是主入口，也从视觉上将露天平台与展示区结合起来。

大部分的窗户都是七巧板一样的几何形，这是为了强调位于主体建筑正前方的起居室拥有开阔的视野，这个房间的一部分由钢结构建成，从展示区延伸至后花园。客人们可以从与他们购买的房子相同的窗户里瞥到这里的全景。

Qiu Chunrui Design Studio

邱春瑞设计师事务所

邱春瑞

Qiu Chunrui Design Studio set up in 2005, focus on providing interior design and soft outfit display supporting to well-known property developers and hotel investment and developers. Design area including hotel, sales centre, model houses, villa, club, restaurant and public space.

Over the past five years, we are known by a lot of people through our works. We have done many different types of space design in a quite short time. And these projects achieved great appraisal and compliment. We accumulated lots of experiences during our cooperation with developers. While strengthen ourselves, we also actively working with others in order to establish the link between interior design and other related industries and to extend our design concept to a wider area.

Spend everyday life easily and do design seriously. We always stick to the principal that 'tailor for each client'. And that's exactly our core belief.

　　邱春瑞设计师事务所成立于2005年，专职为知名地产商和酒店投资开发商提供室内设计及软装陈设配套。设计项目涵盖酒店、楼盘营销中心、样板房、别墅豪宅、会所、餐厅及公共空间。
　　公司成立五年来，我们以作品为载体迅速知名于全国，在短短的时间里，成功设计了全国多个一线城市的房地产开发商的各种类型风格的空间，而这些作品在业内和项目当地都受到很高的评价和赞扬。在与各大发展商的合作中得到了更多的发展空间，累积了丰厚的经验。公司在增强自身实力的同时，积极和其他行业合作，不断建立与室内设计相关行业之通路，让公司的设计理念伸展至更宽更广的天地。
　　轻松做人，认真设计，我们始终坚持"为每一个志同道合的业主量身定做来设计"。这是事务所全体员工于室内设计事业上一直在坚持、也将持续的风格，也正是邱春瑞设计优势的核心理念。

All Love in Town Sales Centre

Design Agency: Qiu Chunrui Design Studio
Location: Shenzhen
Area:1,500m²
Photography: Lv Rongde
Main Materials: bulb, stainless steel ball,
chrome ball, table tennis, stainless steel plate,
stone, veneer, grey mirror

合正汇一城营销展示中心

设计公司：邱瑞春设计师事务所
项目地点：深圳
项目面积：1500m²
摄 影 师：吕荣德
主要材料：灯泡、不锈钢球、镀铬球、乒乓球、
不锈钢板、石材、饰面板、灰镜

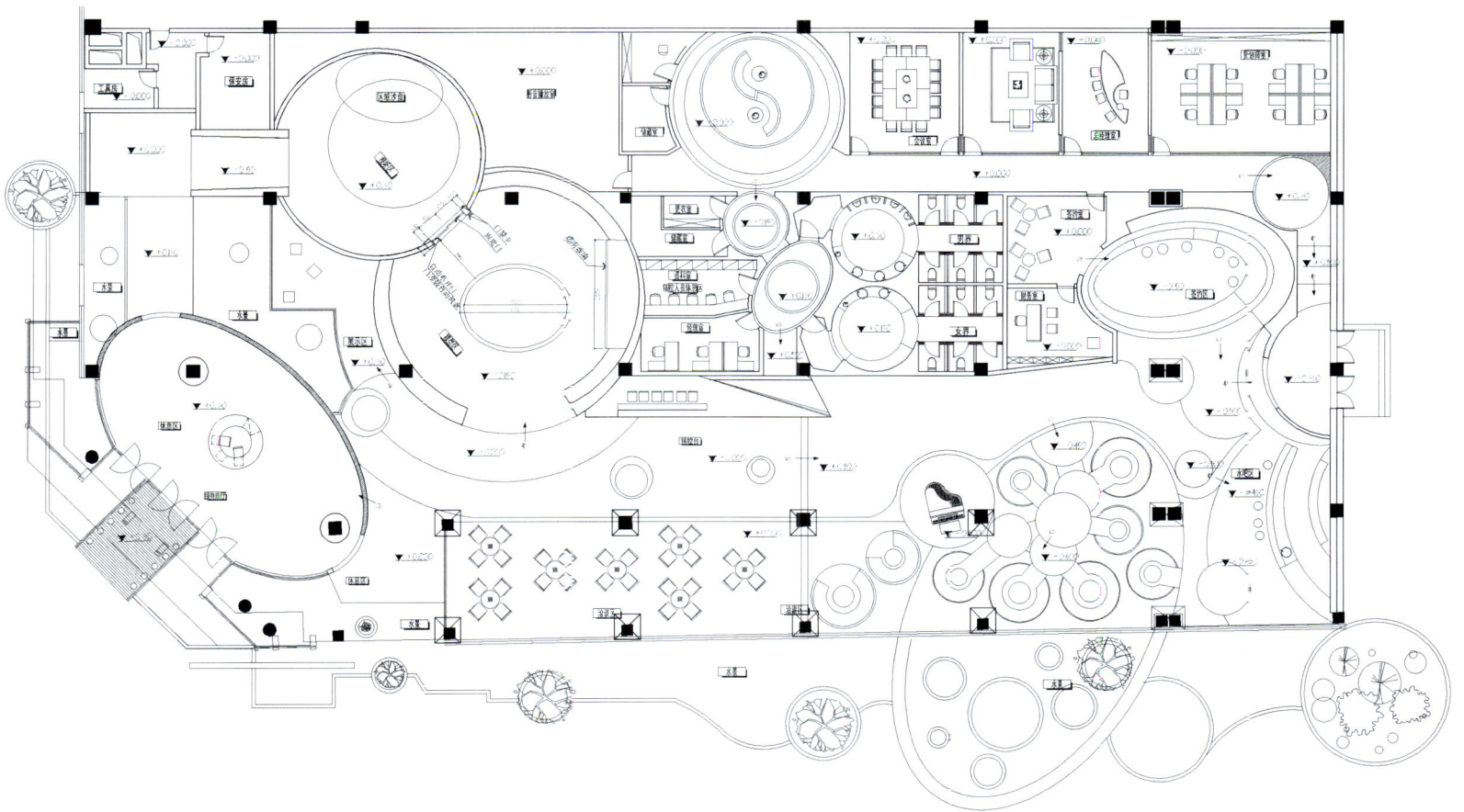

平面图

The entire sales centre consists of reception hall, exhibition zone, negotiation zone and office zone. Entering the door, there is a large-scale LOGO wall that made up of 10,800 stainless steel balls and 25,000 chrome balls, giving people a strong visual impact, powerful and striking.

Designer sets two walkways in the reception area, as the transition area of connecting the other two areas. The right walkway is for clients who come here more than once, walking along with the generatrix into the negotiation area. While the left walkway is an exhibition generatrix prepared for clients who come here for the first time.

When it comes to the two-storey negotiation hall which consists of 130 thousand transparent bulbs, visitors' view will be widen all of a sudden. Designer uses the same design element, the glass lamps as decoration materials in this space.

In the negotiation area, clients can sit on the sponge couch, listening to some melodious music while consulting with professional staff. A bar is allocated at the end of the hall, providing visitors with beverages and cakes. The signing area and the negotiation hall are separated clearly, greatly ensured the privacy and safety of signing and payment. Many vanguard designs in this case have already become the new vane of local sales centre.

整体场域分为四大部分：接待大厅、展示区、洽谈大厅和办公区。由大门进入接待区，迎面而立的是由10800个不锈钢球和25000个镀铬球组成的大型标志墙，给人强大的视觉冲击力，大气而醒目。

在接待区内，设计师构置两条左右动线的走道，作为连接两大场域的过渡区。右侧走道是为再次前来的客户准备的路线，可以顺应动线进入洽谈大厅。左侧走道则是为第一次来咨询的客户安排的参观展示动线。

来到由13万个透明灯泡组成的两层高的洽谈大厅，视野豁然开朗，气势摄人心魄。整体的空间规划，设计师运用相同的设计元素，使用数个晶莹剔透的玻璃灯做装饰材料，融入整个空间的墙面和天花之中。

进到内部洽谈区，客人可以坐在沙发上，伴随悠扬的乐曲，轻松地与专员咨询。大厅底端则配有水吧台，为访客提供饮品和糕点。而签约区与洽谈大厅清楚划分，保障签约付款的私密与安全。本案的多项前卫设计已成为当地售楼处的新指标。

合正汇一城
ALL LOVE IN TOWN

Panshine Environment Art Design Co.,Ltd

派尚环境艺术设计有限公司

李益中

Graduated from the Department of Architecture of
Dalian University of Technology
Founded Shenzhen Panshine Environment Art Design
Co., Ltd. in 1998
Director of China Institute of Interior Design (CIID)
Held solo show of his works in Hexiangning Art
Museum in 2002
Best Interior Designer in 2002
Published *Life of Sample* in 2007
Planed the first Ten People·Space Design Competition
in Shenzhen in 2007
Invited to "APSDA Asia-Pacific Space Designers
Association" and won the Best Design Award in Asia-
Pacific region in 2008
Published *From B to A: Sales Centre Design Strategy*
and hired as visiting professor in Shenzhen Polytechnic
in 2009
Hired as practice mentor of Academy of Arts & Design,
Tsinghua University and School of Architecture, Central
Academy of Fine Arts, etc. in 2010
Currently studying a master's degree on Design
Management in Politecnico di Milano, Italy
He attaches great importance to the design strategy and
the connection between human and environment, the
projects are very conceptual and perceptual, controlled
and elegant

毕业于大连理工大学建筑系
1998年创立深圳市派尚环境艺术设计有限公司
中国建筑学会室内设计分会（全国）理事
2002年在何香凝美术馆举办个人作品展
2002年度中国最佳室内设计师
2007年出版图书《样板生活》
2007年策划首届深圳十人·空间设计竞赛
2008年受邀参加"APSDA亚太空间设计师联合会"并荣
获APSDA亚太地区最佳设计作品大奖
2009年出版《FROM B TO A 售楼处设计策略》，并受聘
为深圳职业技术学院客座教授
2010年，受聘为清华大学美术学院、中央美术学院建筑学
院等艺术院校的设计实践导师
现正攻读意大利米兰理工大学设计管理硕士学位
李益中重视设计的策略性，关注人与环境的关系，作品理
性与感性兼具，节制而优雅

Sales Office of Wuhan Fuxing International City

Design Agency: Panshine Environment Art Design
Co., Ltd
Designer: Li Yizhong, Zhou Weidong
Location Wuhan
Area: 1,200m²
Main Materials: flame retardant metal sand,
dark grey fluorocarbon resin paint, dark red
fluorocarbon resin paint, fiberglass stamp, white
hone finished brick, dark grey hone finished
brick, black shining brick

武汉福星国际城售楼处

设计公司：派尚环境艺术设计有限公司
设 计 师：李益中、周伟栋
项目地点：武汉
项目面积：1200m²
主要材料：阻燃金属砂、深灰色氟碳漆、暗红色氟碳
漆、玻璃丝印图案、白色哑光砖、深灰色哑光砖、黑
色高亮砖

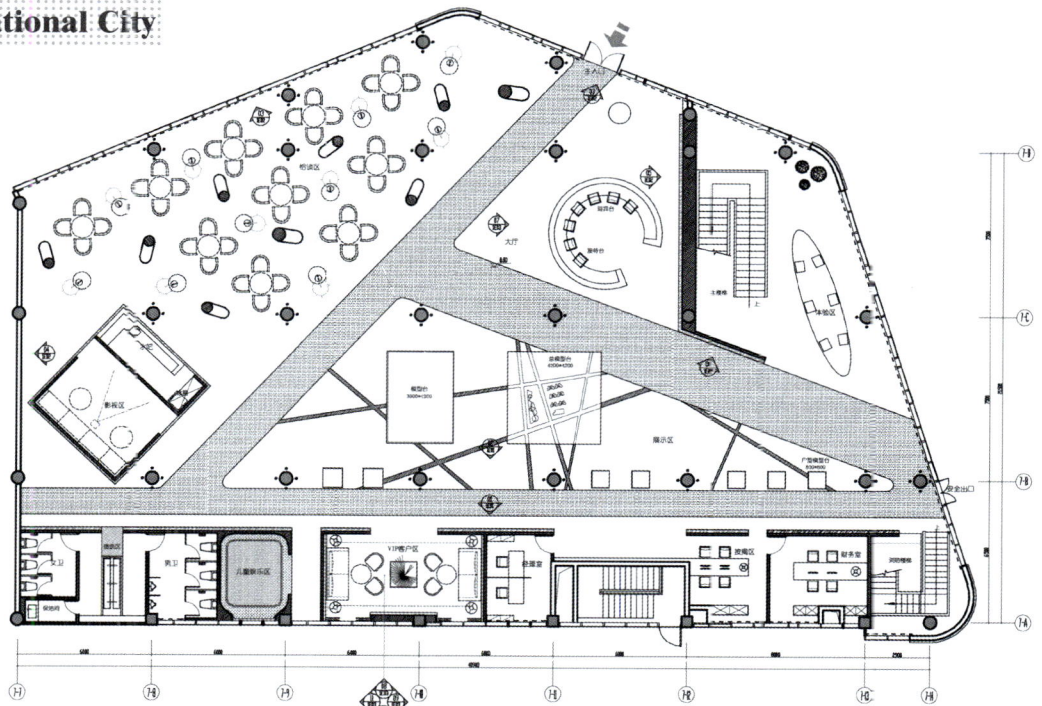

平面图

The design aim of this case is to present a fashionable youthful leisure commercial space, corresponding to the comprehensive plan of Fuxing International City.

In this space, each geometrical element is both perfectly syncretic and independent. The strong contrast of black, white, red and yellow makes a special spacial experience. The oversized luminous ceiling, with a relatively low height, strengthens impressions of space and also separates each functional area, which is an important floodlight source making a soft and modern space tone. The design of the bent column of white light in the discussion area breaks the dreary atmosphere, with the beauty of the black upright column of the original building.

本案的概念设计定位为年轻活力、时尚休闲的商业化空间，与福星国际城整体规划理念相一致。

在这个空间里，各种几何元素既完美融合又各自独立，黑白红黄强烈的色彩对比碰撞出独特的空间感受。超大尺度的发光天花造型，以相对较低的标高，强化了交通空间，并分隔出不同的功能区域，同时也是整个销售中心重要的泛光源，塑造了柔和现代的空间基调。洽谈区白色发光斜立柱的设置，与原建筑黑色直立柱交相辉映，打破了空间的沉闷气氛。

Sales Office of Huizhou Zhongxin Waterfront City Phase 1

Design Agency: Panshine Environment Art Design Co., Ltd
Designer: Li Yizhong, Fan Yihua
Location: Huizhou
Area: 1,800m²
Main Materials: baked lacquer board, artificial stone, acrylic board, natural stone

惠州中信 · 水岸城项目一期售楼处

设计公司：派尚环境艺术设计有限公司
设 计 师：李益中、范宜华
项目地点：惠州
项目面积：1800m²
主要材料：烤漆板、人造石、亚克力板、天然石材

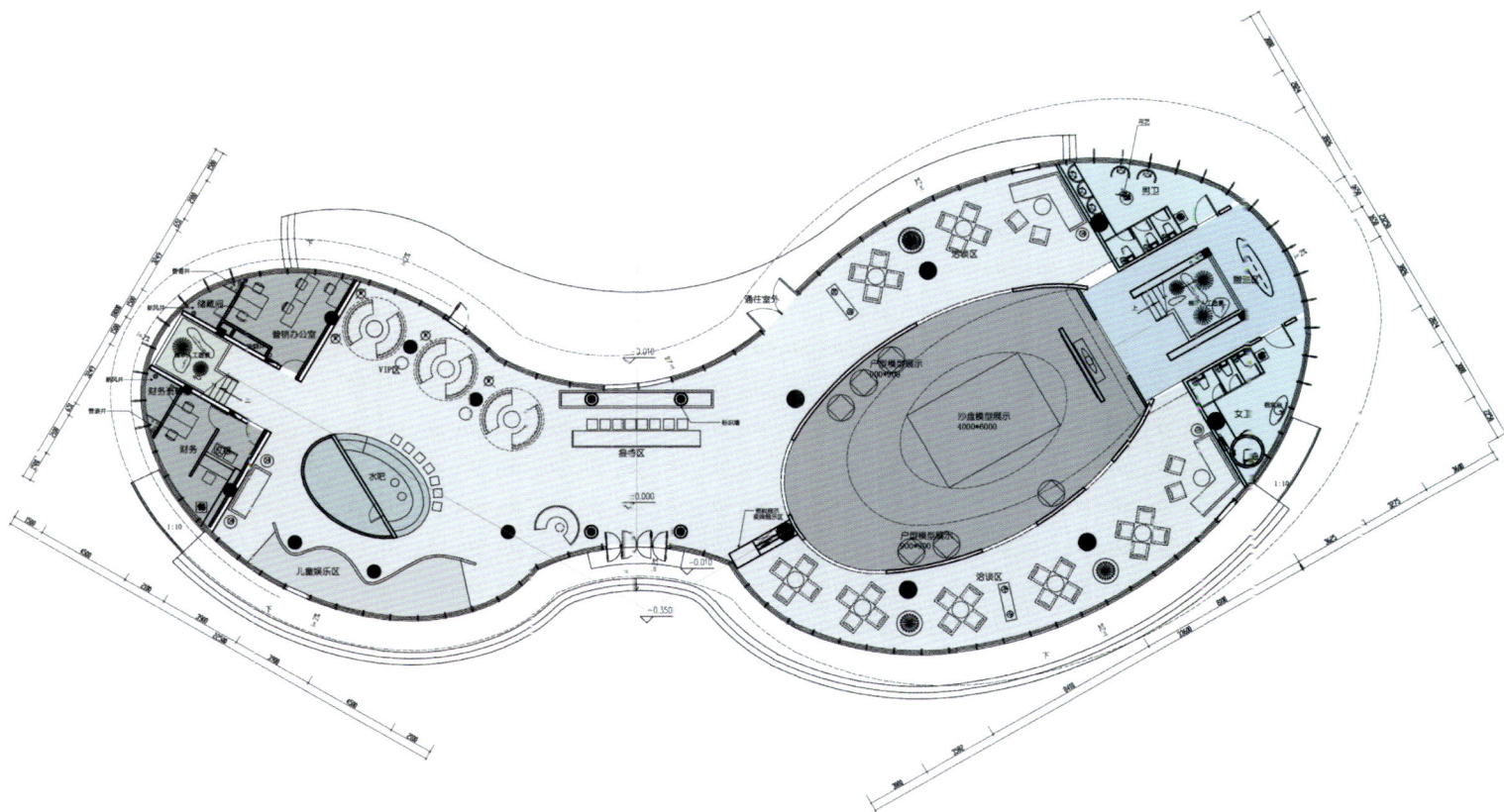

1F平面图

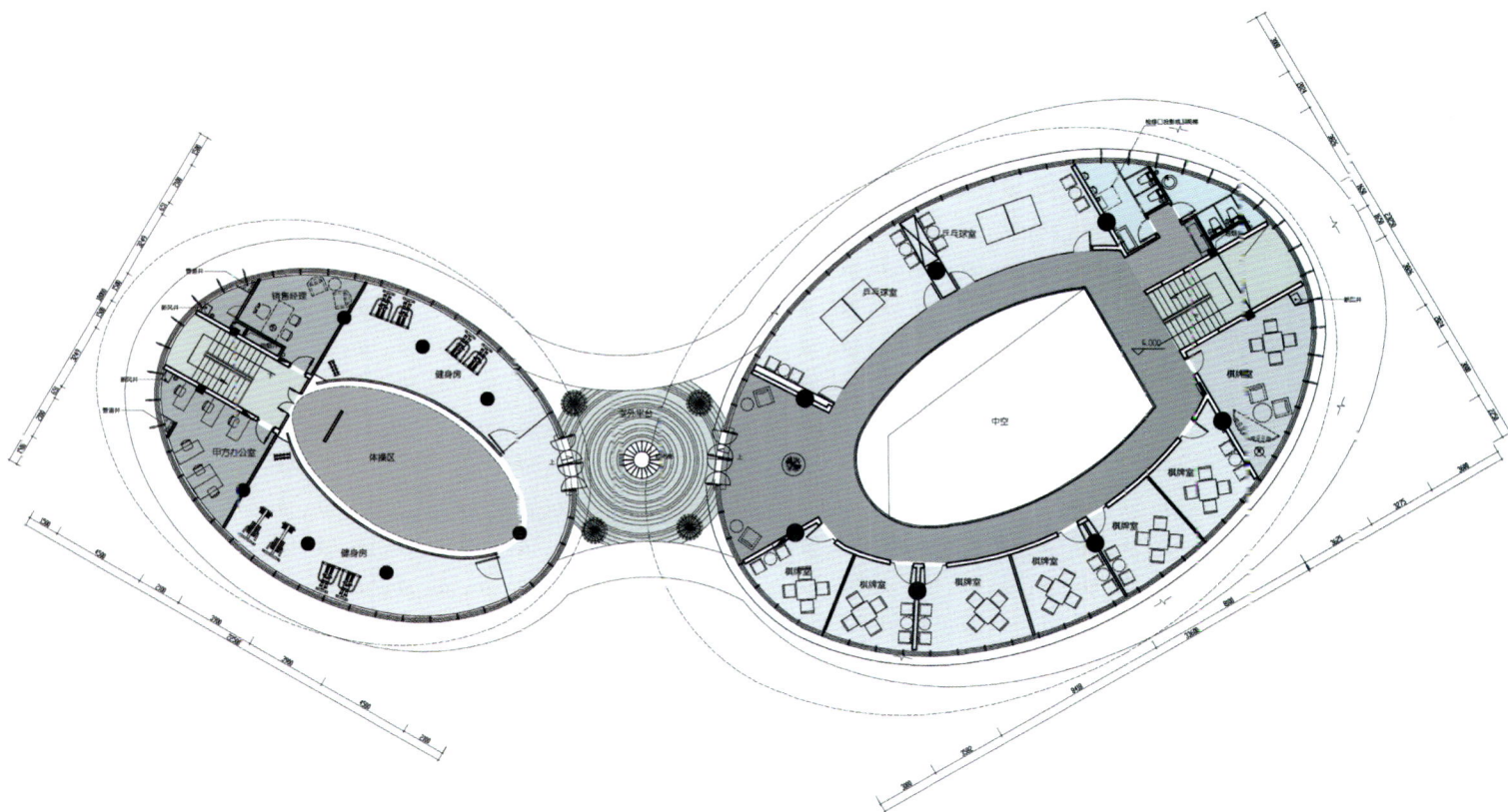

2F平面图

This building, located by the artificial pool, is formed by the two connected oval blocks. The white streamlining building forms a delightful contrast with the waterscape, which gives us a modern and airy visual experience. As to the material, there only added a white external wall to structure layers of floors, besides a transparent glass facade. All the design elements are pure enough. And designers hope to express the most vivid artistic conception in a purest way.

The design aim of this case is to use the simple neoclassical style, experssing a modern international quality free from vulgarity together with the reservation of elegant classical elements. Thus, in the selection of furniture, redesigned classical ones are widely used. And as to spaces colors, black and beige are also chosen to use besides white.

The ceiling is brief but grand. White ellipses are range upon range, which is of white dune-like texture, beautiful and full of the sense of layers especially with the hidden lights.The pilotis design of the courtyard enhances the depth feeling of the space, making the first and second floors share the sufficient lighting from the ceiling. The huge fantastic crystal light on the ceiling of the second floor shines over the sand table area in the first floor, becoming the focus on vision. At this time, people in the house will compulsively dace with a Waltz.

Marbles are put together to form exquisite patterns. The classical black-and-white combination presents a modern feeling, rightly separates the circular sand table area from the negotiation area. There are sketchy figures in the white baked lacquer board on walls, which enrich veins in space. Beige leathers are used for podetium and some partial walls, softening the space as well as improving the quality of space.

However, black is widely use in the reception centre instead of white. The black whole-steel surface of the reception desk is in a brief design, combining well with

black marble floor. Pure colored artistic boards are used for the background wall, showing an effect of ink painting with lighting of lamps.

　　售楼处建筑由两个椭圆形体块连接而成，坐落于人工湖旁。白色的流线形建筑和紧邻的水景虚实相映，给人以现代、轻盈、飘逸的视觉感受。在材质上，除了透明的玻璃幕墙，只加入了楼板结构层的白色外墙材料，色彩上也只有轻盈的玻璃本色和白色。整个售楼处的设计元素极其纯净。因此，设计师希望在室内设计中也延续这种轻盈飘逸的特质，尽可能用最单纯的手法来传达最生动的意境。

　　本案的设计意在运用简洁化的新古典语汇，融合现代建筑形体，表现清丽脱俗的现代国际气质，同时保有优雅的古典元素。因此，在家具的选型上，经过再设计的具有古典韵味的家具被大量选用。在空间色彩上，除了大量的白色，设计师也选用了与之完美搭配的黑色以及中性的米色。

　　天花的处理简洁而大气，椭圆形的白色界面层层重叠，在暗藏光源的作用下，呈现出白色沙丘般的肌理，纯美而富有层次感。中庭的挑空设计，增加了空间的纵深感，也使得一二层共享了天花充足的光源。梦幻璀璨的巨型水晶吊灯通过中庭由二层天花垂至一楼沙盘区上方，成为空间的视觉焦点，也为整个空间带来优雅的古典气质。此时若奏响一只圆舞曲，会让置身其中的人情不自禁的翩翩起舞。

　　地面由大理石构成细腻的拼花，经典的黑与白碰撞出独特的时尚感，也同时界定出圆形的沙盘展示区与环形的洽谈区。墙面白色烤漆板有着形态简约的花纹，在售楼处整体光洁的界面基础上，丰富了空间肌理。柱体和部分墙面饰有米色的皮革软包，柔化空间硬度的同时又提升了空间的品质感。

　　接待处则收敛了白色的应用，以大量黑色铺陈空间。黑色镜钢面的接待台造型简洁，与黑色大理石拼花地面融为一体。背景墙选用纯色系的艺术板，在背投灯光的照射下，显示出泼墨画一般的效果，自然而独特。

Hengji Changsha Kaixuanmeng Sales Office

Design Agency: Panshine Enviroment Art Design Co.,
Ltd
Deaigner: Li Yizhong, Fan Yihua
Location: Changsha
Area: 1,200m²
Main Materials: white latex paint, black specular
stone, noble grey marble, white fluorocarbon resin
paint, custom made black carpet, Chinese black
marble pattern, aluminum paint, stainless steel, black
glass

恒基长沙凯旋门售楼处

设计公司：派尚环境艺术设计有限公司
设 计 师：李益中、范宜华
项目地点：长沙
项目面积：1200m²
主要材料：米白色乳胶漆、黑色高亮光石材、贵族灰大
理石、白色氟碳漆、黑色定制图案地毯、中国黑大理石
图案、铝板烤漆、不锈钢、黑玻

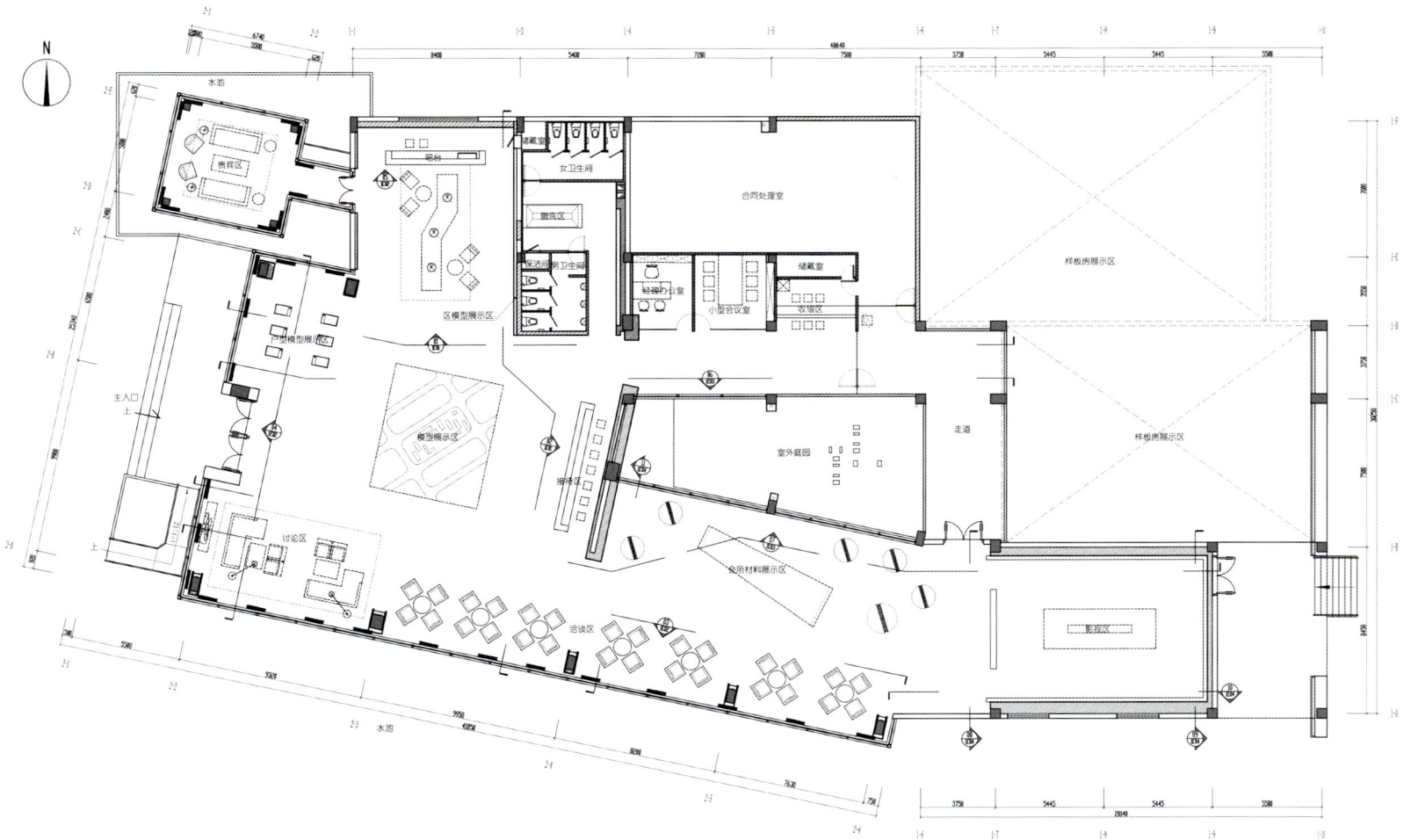

平面图

The aim of the design is to make a modern space both concise and classical. With use of black specular stones, the aluminum paint, stainless steel, black glass etc, it rightly coordinates with the outsides of the building and the overall positioning of the project, expressing a grand and modern space. At the same time, the selection of furniture and decorations bring a classical feeling, such as the tea table in the negotiation area and the model table, which is suit to the noble image of this project.

The logo of the project, as a sign language, shows the quality and style of the whole real estate, which is a big feature of the project. The whole ceiling and floor of the model display area and the reception together with the background wall of the reception are arranged with the real estate logo pattern, which highlight the visual impact and also strengthen clients' impressions.

本案的设计定位为简洁干练又蕴含古典韵味的现代空间。设计师运用简洁的线条、黑色高亮光石材、铝板烤漆、不锈钢、黑玻等整体刻画出一个大气、现代的空间格局，与空间交错叠加的建筑外观以及开发商对楼盘的整体定位相呼应。而空间内部家具与配饰的选择，比如，洽谈区茶几、模型台等又点缀出一些古典的韵味，正切合凯旋门显赫尊贵的概念意象。

楼盘标志作为楼盘特征的一个标识性语言，体现着整个楼盘的风格和气质。设计师巧妙地将这一设计语言运用到空间中，成为本案的一大亮点所在。在主入口处的模型展示区及接待台、整个天花造型、地面以及接待台背景墙，全部使用楼盘标志排列组合成的图案，极富视觉冲击力，突显了售楼处的气质，从营销角度来看，又加深了客户对楼盘的印象。

Sales Centre of Shenzhen Gaofa the 5th Avenue Phase 3

Design Agency: Panshine Enviroment Art Design Co., Ltd
Designer: Li Yizhong, Fan Yihua
Location: Shenzhen
Area: 922m²
Main Materials: dark grey lacewood, striped carpet, grey mirror, grey steel, dark grey baked lacquer board

深圳高发第五大道3期营销中心

设计公司：派尚环境艺术设计有限公司
设 计 师：李益中、范宜华
项目地点：深圳
项目面积：922m²
主要材料：深灰色尼斯木、条纹地毯、灰镜、灰钢、
深灰色烤漆板

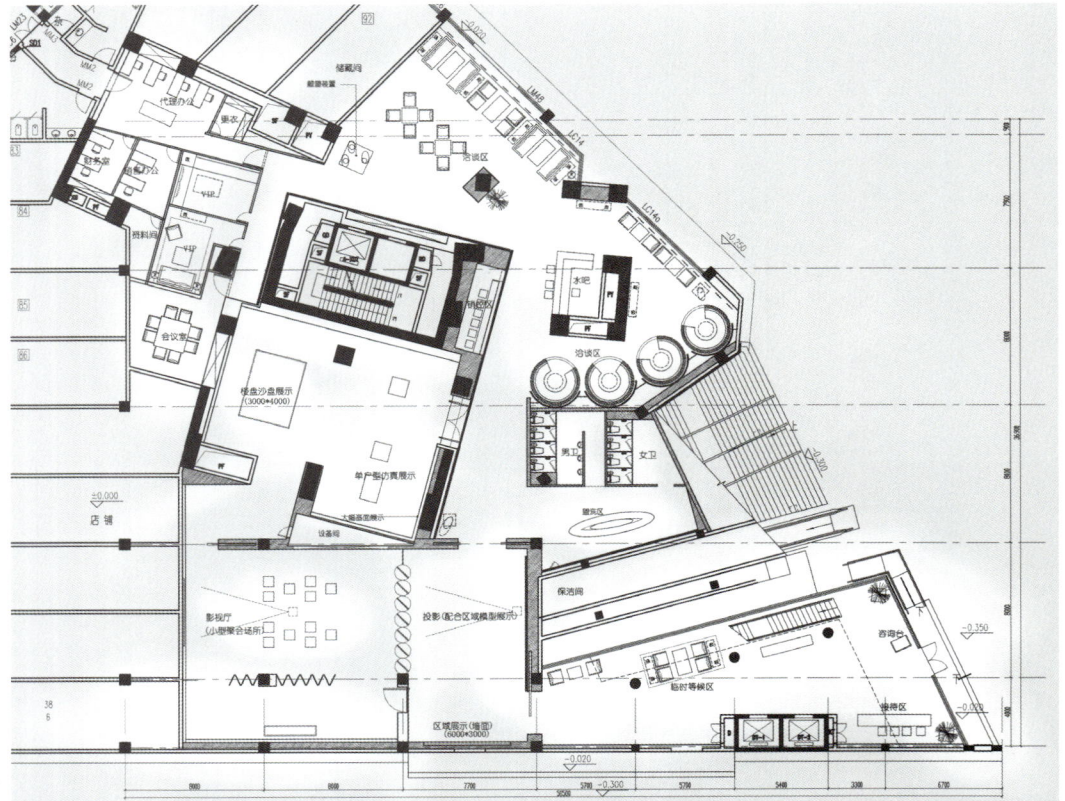

平面图

Light is the main figure of this case. There is a special order from the quiet shadowy exhibition hall and the open and bright discussion area. The light guidance and the gradual change of color give underlying visual and psychological guidance, which makes the curious people go to find the end part of the sales house. A quiet and mysterious atmosphere is well presented here with the refined and reconstructed Eastern elements designed by the designer.

The design of this case has made full use of the features of the original construction which is fragmented because of too many structural column. Thus, the design is longer limited to the traditional space settings of the sales centre, instead, it skifully uses this special character. Each functional area is well separated. And with the use of doors, they are also in good connection, making a spacial sense of rhythm with changes with lights and colors. Visitors will feel like wandering in a Chinese garden which is deep and quiet, meanwhile, they will visit all the sales centre inadvertently.

The oriental aroma of serenity and harmony is what designers really want. Still in the design process, designers not only make good use of the essence of Chinese culture but also blend it with modern techniques, making a combination of time and space.

The separated spaces are unbroken, extended and interpenetrated, but not too hollow and dull. Each function area is enclosed by screen doors, wooden grating, etc. The discussion area is enclosed by dark grey lacewood bars, presenting a unique space with the circular teapoy of the same material. It separates the negotiation places and enhance the privacy. Still, each negotiation place is not totally enclosed. Through gaps of gratings, the contiguous space can also be detected. There is a stone water bar counter in the centre of the discussion area, like a pavilion in the space. The selection of massive furniture and lights of primitive simplicity strengthen the rich oriental flavor. And the use of wall picture improves the qualities. The aisle wall is a well-designed relief from audiovisual area to the discussion area. And the brief white china plate in the centre of the relief presents the Zen feeling of the kare-sansui, which seems static in time under the lighting.

　　光是本案一条结构清晰的脉络，静谧幽暗的展示厅和开阔明亮的洽谈区之间有着特殊的空间礼序，灯光的引导和色调的渐变给人以潜在的视觉和心理引导。使人流在好奇心的驱使下，一步步寻往售楼处流线的终端。东方元素经过设计师的提炼与重构，营造出宁静而又神秘的氛围。

　　本案是在原有售楼处的基础上改建而成的。原建筑中存在太多承重柱，使空间支离破碎，难以整合。基于此考虑，设计师不再局限于传统售楼处流畅、直白的路线设置，而是巧妙地利用空间这一独特之处，使各功能区各自独立。同时利用门禁的方式将其有机的串联在一起，结合光影和色调的变化，产生忽开忽合、时收时放的空间节奏感。客户游走其间，感受到的是犹如中式园林一般的深邃藏幽的情趣，同时也在不经意中，循序渐进地探访完销售流线的全程。

　　宁静和谐的东方韵味是设计师想要在空间中表达的。在这一过程中，设计师并不限于对东方元素的堆砌，而是吸取东方文化精髓，利用现代的手法进行再创造，使两者之间形成一种跨时空的结合。

　　被分隔的空间多相互连绵、延伸、渗透，而不流于空旷、单调。设计师提取中式建筑语汇，利用屏风门、木格栅等传统中式元素围合各个功能区。洽谈区利用深灰色尼斯木木条圈合，配以同材质圆形茶几，形成独特的空间的造型，同时也将各个洽谈位分隔开来，增加了私密性。洽谈位之间也并非全然封闭，透过木格栅的缝隙，可以隐约窥见相邻空间的动态，整体营造出一种似隔非隔，似断非断的宁静空间氛围。洽谈区中央利用石材围合成一座水吧台，仿若玉立于空间中的一座亭阁。古朴厚重的中式家具与灯具的使用，深化了空间浓厚的东方气息；挂画的运用，则在整体上提升了空间的气质。影视区过渡到洽谈区的走廊墙面，是艺术家根据空间气质精心创作的雕塑画，形态简约大方的白色瓷盘悬立于画中央，如同枯山水一般引发禅思，在灯光的映照下，时间似乎在一瞬间凝滞。

Taipei Base Design Centre

台北基础设计

Janus

Roy

Top 10 Most Popular Designers
Top 100 Interior Design Company selected by *IDChina*
Most Powerful Designers in China
The winner of Red Dot Award Communication Design 2007
One of the well-known interior designers both in Mainland China and Taiwan China
Was invited to give a speech for *Home Decoration* Salon in Shanghai
The 7th Modern Decoration International Media Prize
The TID Award of Space Furniture
Many entries projects of The TID Award of Residential space
Top 10 Most Popular Designers selected by *IDChina*
IDChina Top 10 Designers of The Year
The Golden Bund Award(Commercial Space/Office Space/ Residential Space)
The TID Award of Residential Space
Top 100 Most Popular designers of Taiwan China
The winner of Taipei Association of Interior Designers Awards (Residential Space/Commercial Space/Public Space)

中国十大最受欢迎设计师奖
美国《室内设计》中文版杂志最强百大室内设计企业
中国最具影响力设计师
德国红点奖2007商业及公共空间类大奖得主
海峡两岸室内设计名家设计师
受邀至上海参与《家饰》杂志美食与设计沙龙演讲
第七届现代装饰国际传媒奖杰出设计师
TID空间家具奖
入围TID住宅空间数奖
美国《室内设计》中文版杂志最受欢迎设计师
美国《室内设计》中文版杂志十大封面人物
金外滩数奖(商业空间/办公空间/居住空间)
TID住宅空间奖
台湾百大设计师
TAID奖(住宅空间/商业空间/公共空间)

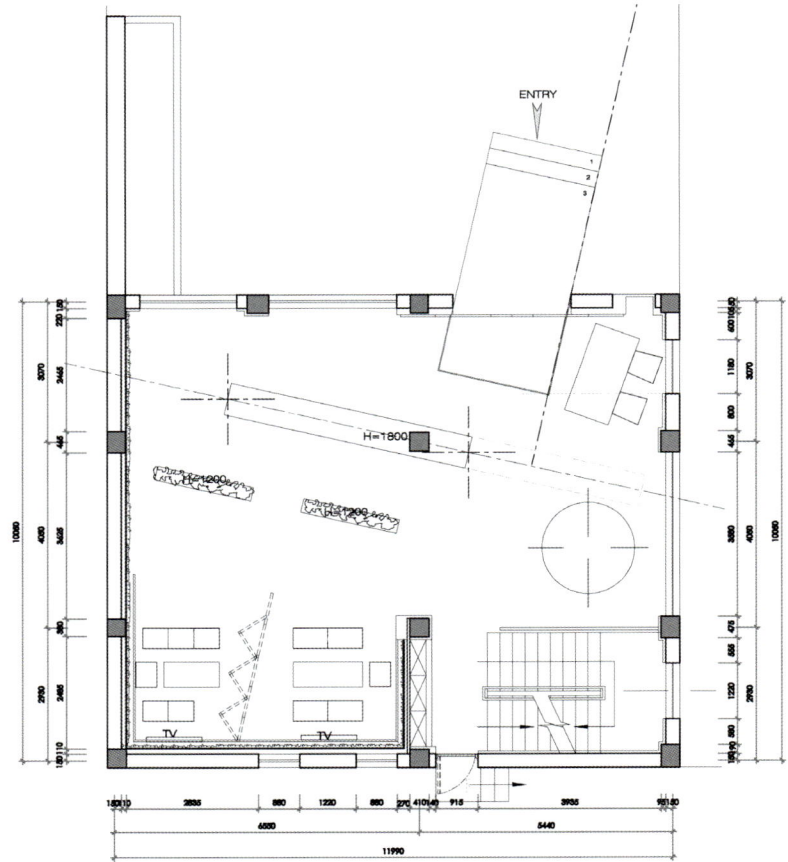

1F平面图

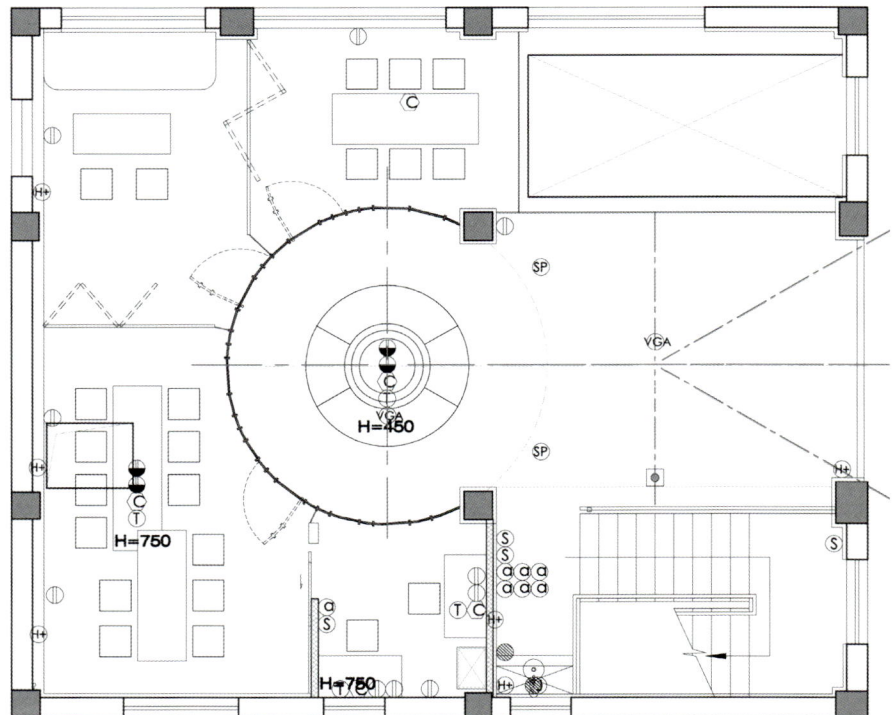

2F平面图

Hua House in Shanghai

Design Agency: Taipei Base Design Centre
Location: Shanghai
Area: 500㎡
Photography: Black Wong
Main Materials: thick transparent toughened glass, grey mirror, silver mirror, tawny mirror, leather, luminous ceiling, marble, solid wood

上海华屋馆

设计公司：台北基础设计中心
项目地点：上海
项目面积：500㎡
摄影师：Black Wong
主要建材：厚钢化透明玻璃、灰镜、银镜、茶镜、皮革、流明天花、云石、实木

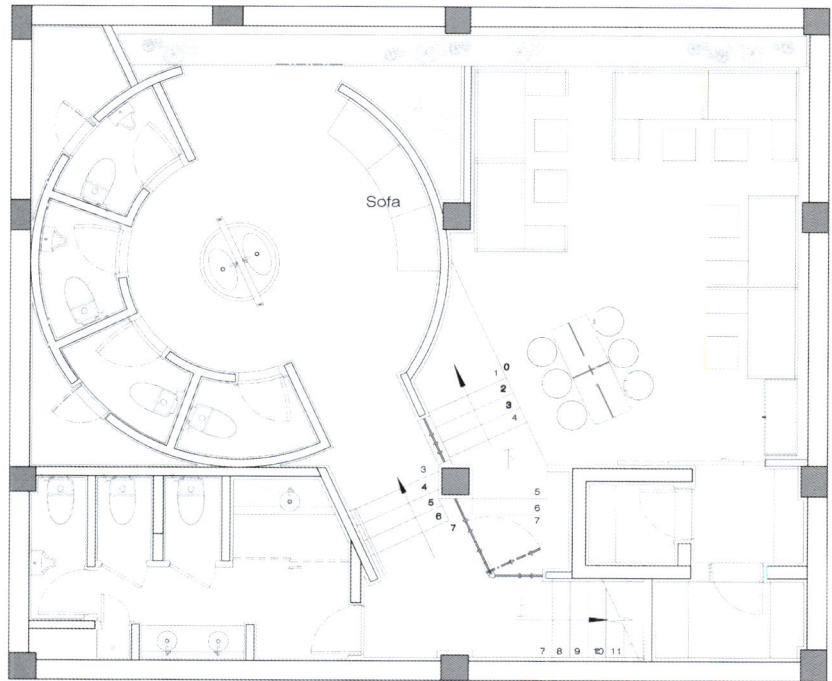

B1平面图

This project is a properties intermediary agent located in Shanghai luxury zone.

Walking along with the generatrix to the model zone, the delicate arc exhibition space and the white curtain hanging down from the ceiling echoed with each other, producing a perfect imagery. This three storeys space was indeed not high enough. How to make visitors feel comfortable in this limited space became one of the key point in the design of this case. The relatively private negotiation zone was where homebuyers may spend the longest time. Hence, designer created the most spacious visual effect without affecting the building structure. The windows extended to acme both vertically and horizontally, letting more light in. Abundant lighting, smooth wood floor, waving greenery outside the windows, all of these together made the visitors feel as comfortable as at home while negotiating.

此案是位于上海豪宅区的房产中介公司。

沿着动线来到模型区，精致的圆弧形展示空间自天花板延伸出白色帷幕罩着灯源，天地双圆呼应，巧妙地传递出圆满的凝聚意象。三层楼的室内空间，其实每层楼的楼高并不高，在有限的楼高中，如何让参观者能感受到最自在无压的氛围，是设计师形塑此案的重点之一。较为私密的洽谈区是购屋者会停留最久的区域，因此设计师将空间释放出最宽敞的视觉效果。在不影响建筑结构的情况下，窗户的垂直与水平皆延伸极致，尽可能将光线揽入室内。充足的采光与温润的木地板交织出温暖的调性。窗外绿意摇曳生姿，让参观者在洽谈的同时，能获得如在自家中的舒适感。

Keiichiro SAKO

迫庆一郎

2004 – 2005 Visiting Scholar at Columbia University.
            Overseas artist dispatched by Agency for
            Cultural Affairs.
2004   Established SAKO Architects
            Jointly presided over Asian Architects
            Associates
1996 – 2004 Worked with Riken Yamamoto & Field
            Shop
1996   Obtained a Master's Degree from the
            Tokyo Institute of Technology
1994   Graduated from Tokyo Institute of
            Technology
1970   Born in Fukuoka, Japan

Awards
2010   2009-2010 Annual International Design
        Award(China)
        Jintang Prize—China Interior Design Awards
        2010(China)
        Honorable Mention Media's Focus

2009   EuroShop RetailDesign Award 2009
        One of Three Best Stores Worldwide :
        ROMANTICISM2 in Hangzhou
        Jin Pan Awards Commercial Sector(China) :
        BUMPS in Beijing , LATTICE in Beijing

1970年   生于日本福冈县
1994年   毕业于东京工业大学
1996年   东京工业大学研究生毕业
1996年~2004年   就职于山本理显设计工场
2004年   成立SAKO建筑设计工社（中国北京）
2004年   主持东方设计公社
2004年~2005年   作为日本文化厅的外派艺术家赴哥伦比
              亚大学担任客座研究员

主要获取奖项
2009年   EuroShop RetailDesign Award 2009
        One of Three Best Stores Worldwide：杭州
        浪漫一身2
        金盘奖年度最佳商业楼盘奖：北京冲击，北京
        格子
2010年   2009~2010年度国际设计艺术成就奖
        金堂奖(中国)
        年度媒体关注提名奖

Pixel in Beijing Modelroom

北京像素样板间

Design Agency: Keiichiro SAKO
Location: Beijing, China
Area: 3,000m²

设计公司：Keiichiro SAKO
项目地点：Beijing, China
项目面积：3,000m²

The modelhouse of high-rise residential building with sales centre to be in accordance with Skip-Floor Dwelling House's buildings for ten thousand households.'Pixel' concept flexibly apply to many parts of interior and exterior decoration. Ceiling filled with paper lightings carry out the impact of the same kind of facilities. Modelhouses divide into four types of different themes such as 'POP'.

The 'pixel' sales centre developed overall 'cell' concept of 'Pixel in Beijing' on the basis of former sales centre. The appearance was formed of many separate tubes accumulated. It is consisting of four kinds of white, light grey, dark gray and black tubes. To overlap and accumulate so as to produce bump sense.

Sales centre hall is a very important part of sales centre used as the whole project sales centre, including negotiation room, guest rooms and so on. The hall, including entrance hall, seeks to pursue a sense of luxury and dynamic space to match negotiation room.

Interior decoration also developed the cell concept. The wall accumulated different length and glossy wooden cases, 400mm×400mm, to get bump and layer sense. The roof is created by 400mm×400mm dropped tubular lightings. Its borders are made of Japanese paper. Then the ground integrated outdoor and indoor, integrated outside landscape to inside.

超高层综合住宅开发楼盘的样板间兼售楼处。与大约一万户跃层式户型组成的住宅设计一致，"像素"灵活运用于外装及内装的多个部分。将吊顶满布由和纸做成的照明灯具，实现了同种设施中所追求的冲击力。样板间分为"流行"等四种不同主题的类型。

北京像素销售中心是在原有销售中心的基础上，重新使用"北京像素项目"整体"细胞"概念这一风格而建成的。外观是由多个各自独立的管堆积成一体构成的。管分为白色、浅灰色、深灰色、黑色四种，各色管长度不一，重叠累积，从而产生凸凹感。

销售中心大厅是用做"北京像素项目"整体销售的场所，包括洽谈室、贵宾接待室等等，是销售中心非常重要的组成部分。包括来宾入口在内，整个大厅设计都追求一种拥有和洽谈室相称的豪华感和震撼力的空间。

内部装修也延续"北京像素项目"整体细胞观念，墙壁部分堆积400mm×400mm的正方形木箱，各个木箱有不同的长度和色泽，创造出凸凹层次感。日本纸构成的400mm×400mm的筒从顶棚上垂下来，起到照明作用。地板部分注意内、外部分的一体性，是将室外园林风格直接引入室内的一种风格。

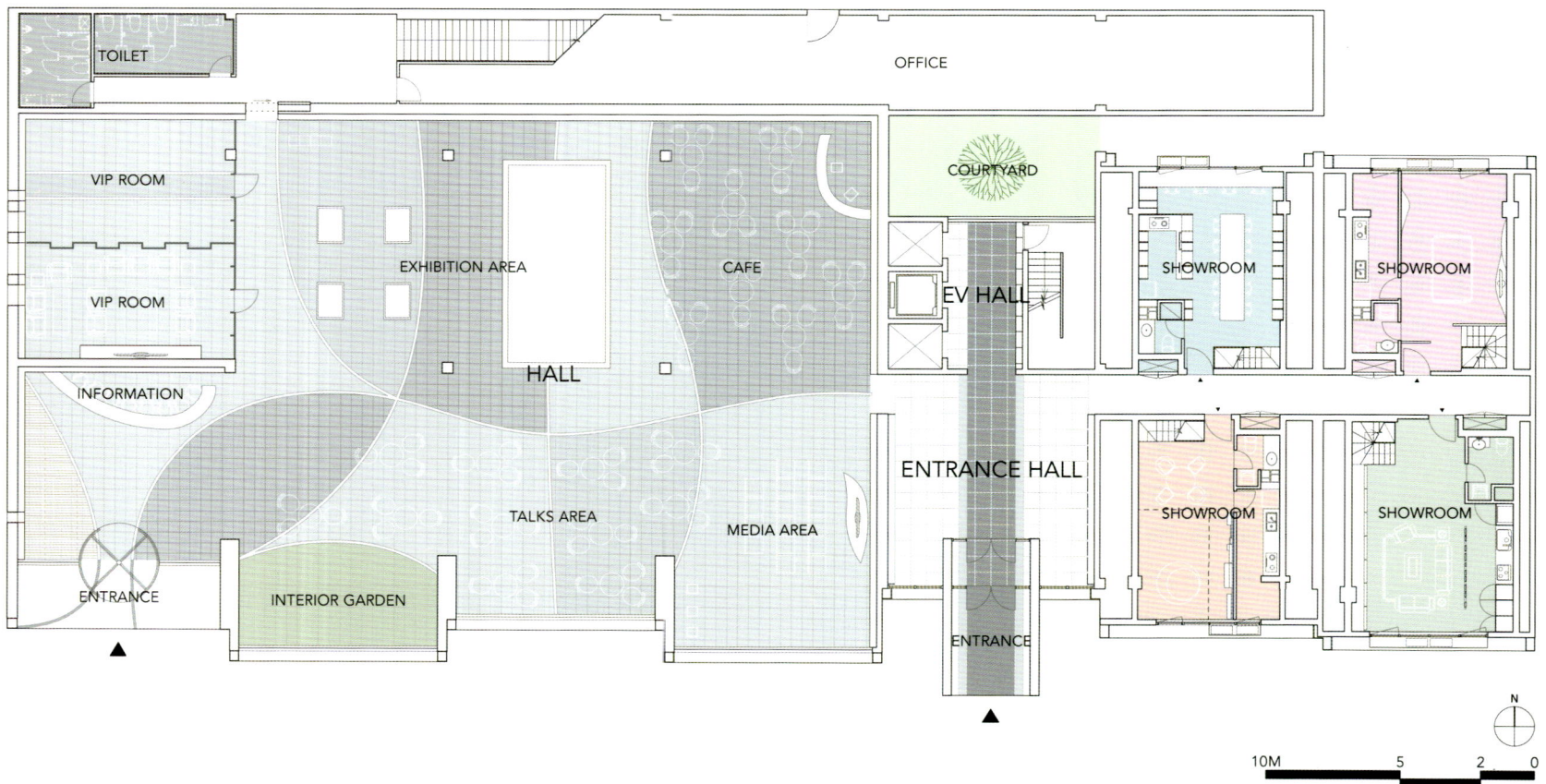

平面图

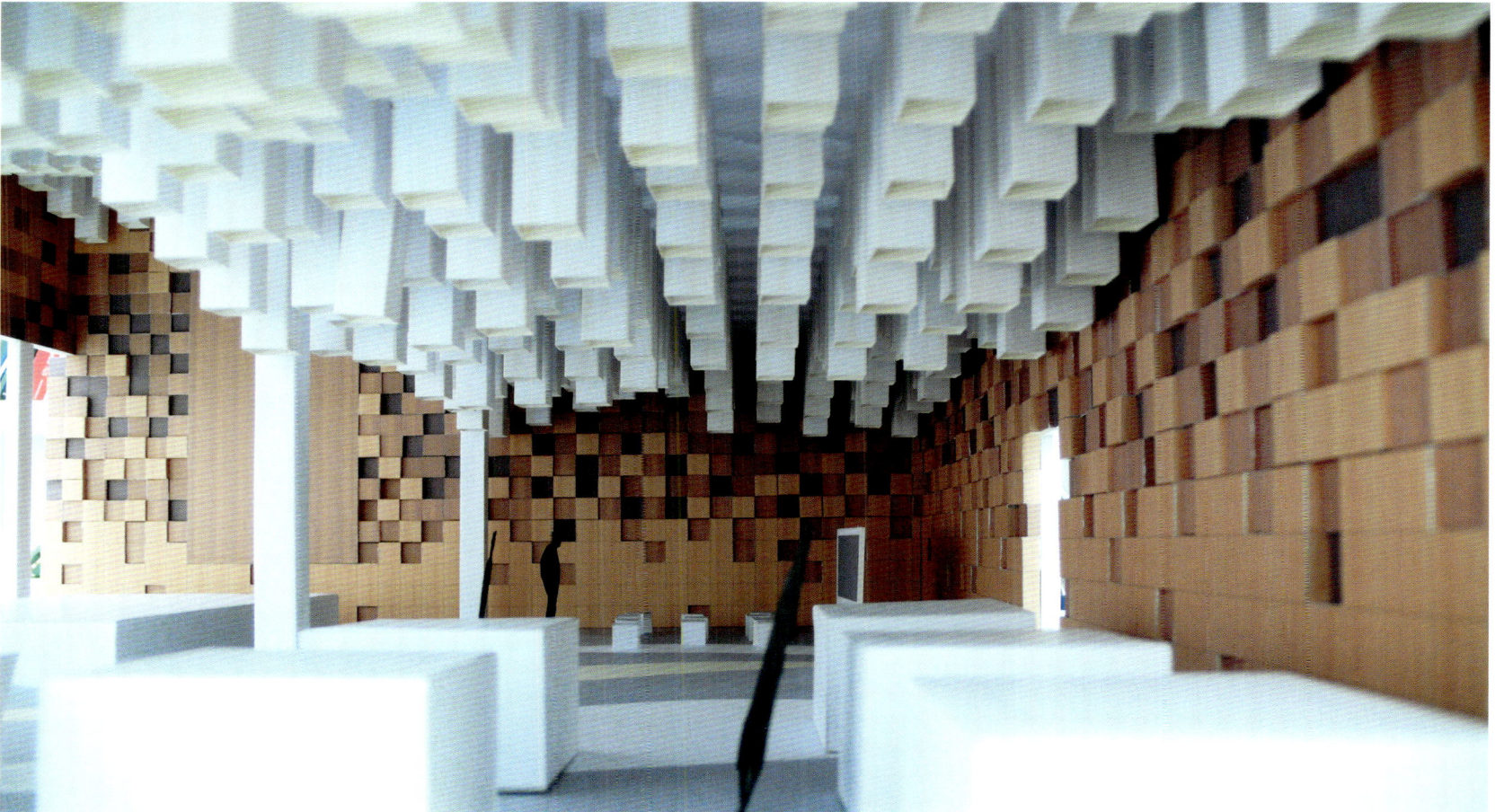

大匀国际空间设计

林宪政

The co-director of Symmetry Design Centre
A lecturer of Taiwan Shih Chien University
One of Taiwanese Top 10 Space Photographer
A professional member of IFI
A professional member of CIDA

Participated projects
2005   Rich Garden in Gubei developed by Shanghai
Urban Construction Group
2005   Crystal Seasons office building developed by
Shanghai Longyuan
2006   Xiaoan Bridge developed by Shanghai Mingyuan
Group
2006   Provence developed by Zhengzhou Sino Land
2006   Class Garden developed by Ningbo Concord
2007   Serenity Coast in Sanya, Hainan
2008   The Life Museum developed by Hangzhou
Younger

大匀国际设计中心协同主持人
中国台湾实践大学讲师
中国台湾10大空间摄影人
IFI国际室内装饰协会专业会员
中国室内装饰协会专业会员

参与
2005 年 上海城建 [上海古北瑞仕花园]
2005 年 上海龙元 [晶采四季办公楼]
2006 年 上海明园 [小安桥]
2006 年 郑州信和 [普罗旺世]
2006 年 宁波和协 [风格城事]
2007 年 海南三亚 [半山半岛]
2008 年 杭州雅戈尔 [The Life Museum]

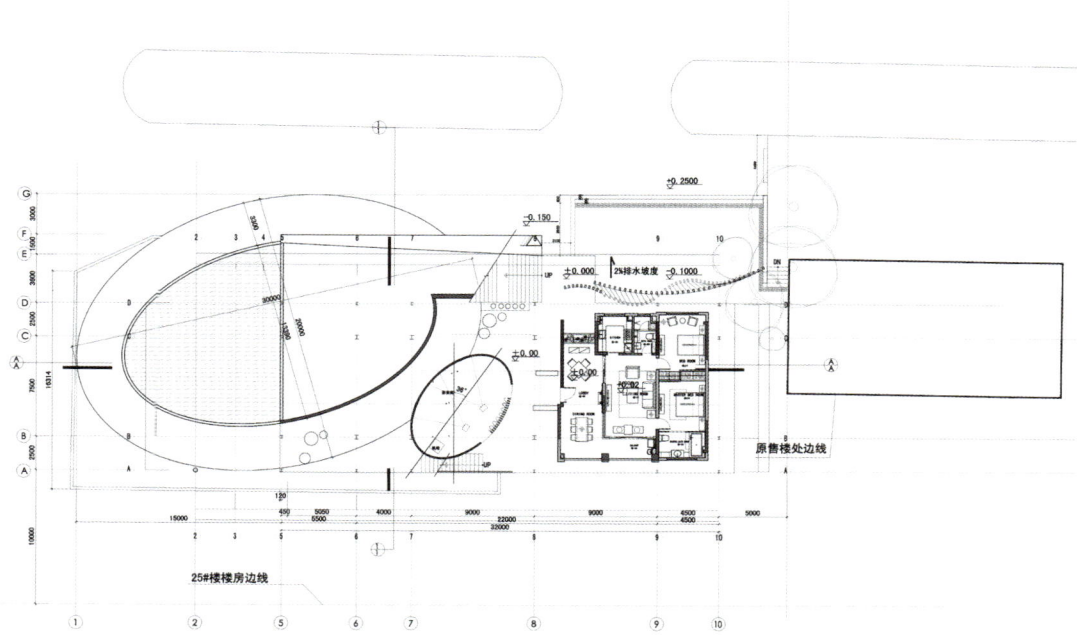

1F平面图

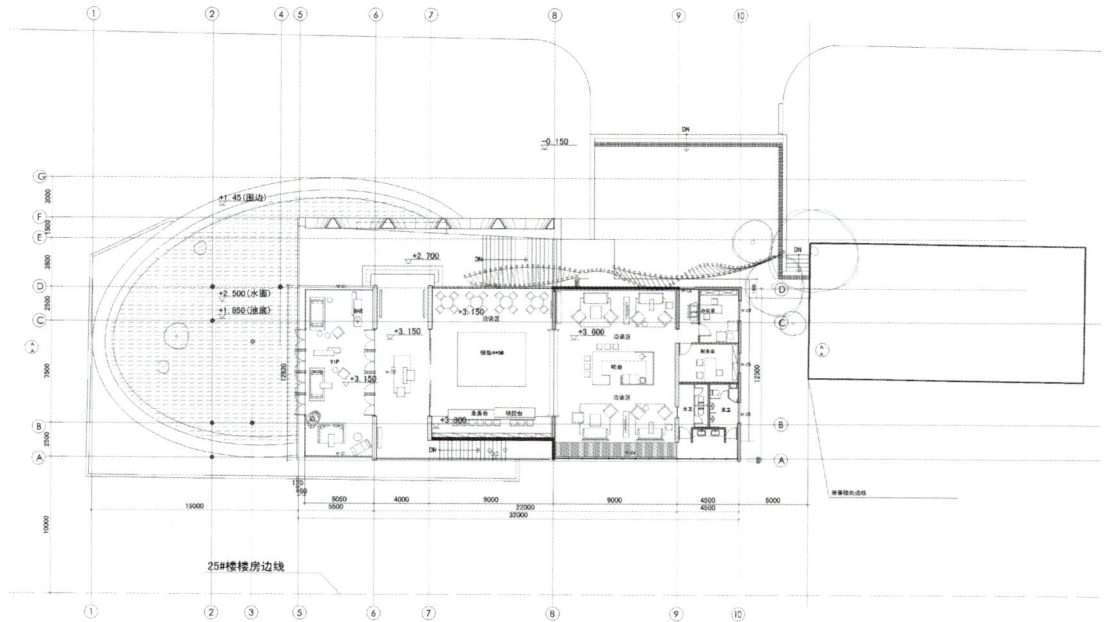

2F平面图

The Reception Club of Sanya Peninsula Blue Bay

Design Agency: Symmetry International Space Design, MoGA
Decoration Design
Location: Sanya, Hainan
Client: Sanya Runfeng Construction Investment Co., Ltd.
Area: 939m²
Main Materials: white exterior wall painted with stucco, black
ceramic brick, crystal jade, and carpet

半岛蓝湾接待会所

设计公司：大匀国际空间设计、MoGA太舍馆装饰设计
项目地址：海南三亚
客户名称：三亚润丰建设投资有限公司
项目面积：939m²
主要材料：白色外墙抹灰、黑色陶砖、透光玉石、地毯

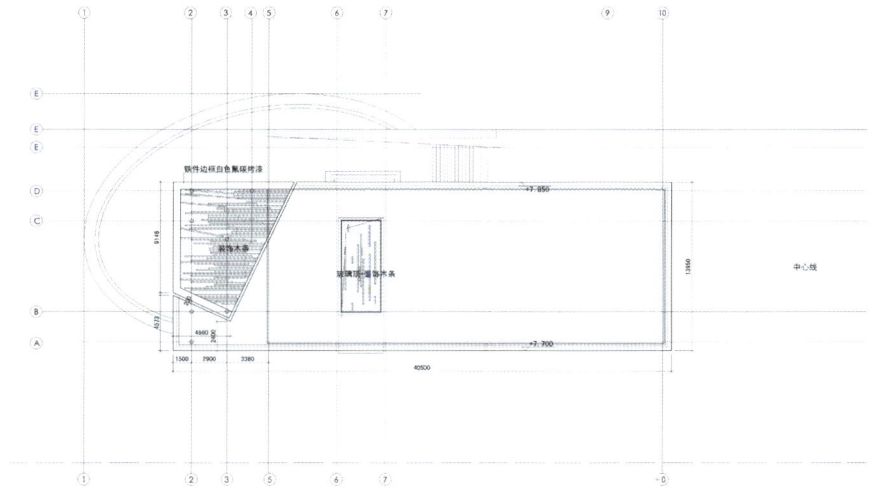

屋顶平面图

Sailing Day And Night, Roaming About Sanya

Located in the inner Yulin Bay, Peninsular Blue Bay faces mountains in its three sides, embracing by Phoenix Hill, Liudao Hill, Mengguo Hill and others, ups and downs. The Reception Club in the distance just resembles a vessel, propping up its white sail and travelling freely at sea. The external surface adopts white tensioned membrane to make something symbolized masts with deep bowl-shaped water at the end. Matching with the waterfront platform of wooden shutter, the whole part looks like the sampan of the bow. Feasting with the harmonious scenery of blue sea and sky, your restless heart will be released from the urban hustle and bustle immediately, and enjoy those nature gifts peacefully. Large employment of floating canopies outside the buildings shelters more sunshine. By combining the physics knowledge with space designing skillfully, the goal of 'enjoying natural shady, reducing energy consumption' has already achieved here.

The Reflection of Natural Beauty

In the morning, breathing the fresh air, your appreciation of the original coastal life starts here; at night, the lighting system of the buildings outlines their whole silhouettes. The decorated details, such as waved wooden grid, huge bowl-shaped pool, and the integral panel with white line board designing, mix harmoniously with the texture and the gloss of the jade decorated wall inside the buildings. With the metal chandelier in champagne color at the roof, it turns into a fantasyland here. Meanwhile, the black tiles in small pieces pave a herringbone-pattern floor, fully demonstrating the unadorned features of the naturally handmade, and visually contrasting to the gorgeous decorated wall.

Delicate Space Designing With Leisurely Boundless View

In the general interior design, the ground floor (1F) is assigned for audiovisual chamber and the simulative sample house; the second floor consists of open meeting area for VIP and the middle corridor in courtyard form, which separates the meeting area with the noisy commercial area. In front of the jade setting, the innermost part of the building, a chandelier of silver foil is decorated here, making the whole bar area as gorgeous as in a yacht, and keeping the leisure feeling at the same time. Walking along the steps, you can go up to the highest where locates a deep and vast pool. Turning inside the building to the middle corridor, you will find the independent & open VIP area and business negotiation area on your both sides. Then going down the steps, the multimedia audiovisual chamber, introduction area for overall planning, and the demonstration area for sample house with fine decorations will turn up within your sight.

昼夜远航，漫步三亚

半岛蓝湾依托榆林内海湾而建，三面环山，围合于凤凰岭、六道岭、孟果岭等连绵起伏群山之间。远望中的接待会所建筑，就像是一艘撑起白帆在海中遨游的船，外立面由白色张拉膜撑起做桅杆，端部配以深邃的碗状水体，迎合木百叶制造出的亲水平台，犹如船头舢板，承接出海天一色的壮景，即刻邀您远离城市燥热，感受大自然的恩赐。建筑物外部大量的飘檐设计，能充足遮阳，真正从物理与空间关系的结合出发，达到"自然阴凉、减少空调能耗"的目标。

自然映射，气韵传神

清晨，呼吸一抹清澈空气，体味原汁海滨生活；夜晚，建筑照明勾勒出整体外廊。波浪形的木格栅、巨型碗状的水池，整体壁板采用白色线板设计，与室内玉石主题墙发散开的纹理光线相呼应，顶部配以香槟色金属吊灯，营造出梦幻氛围，而黑色小瓷砖铺成的人字形地面，一展自然手工的质朴调性，和华丽的主题墙面形成反差式的视觉对话。

精致空间，悠然开阔

在室内整体设计中，地面层1F布置以影音室、模拟样板房；二楼，为开放式贵宾洽谈区，天井式中廊将喧闹的销售区隔开，而建筑最深处的玉石背景前，缀着银箔的大吊灯，为整个吧台区营造出游艇般华丽却不失休闲之味。拾阶而上行至高点是深邃无边的集水池，转入中廊后，左右两侧分别是独立的开放式VIP区和销售洽谈区，再往后，则由楼梯向下进入多媒体影音室、整体规划介绍区和精装修样板间展示区。

The Reception Lounge of Narada & SPA Hotel

Design Agency: Symmetry International Space
Design, MoGA Decoration Design
Location: Sanya Lingshu
Client: Narada Hotel Group
Area: 1,011m²
Major Materials: bamboo chopsticks, collage of
miscellaneous woods, plaster, wood and bamboo
curtains

香水君澜接待会所

设计公司：大匀国际空间设计、MoGA 太舍馆装饰设计
项目地点：三亚陵水
客　　户：香水君澜酒店有限公司
项目面积：1011m²
主要材料：竹筷、拼贴杂木、灰泥砖、木竹帘

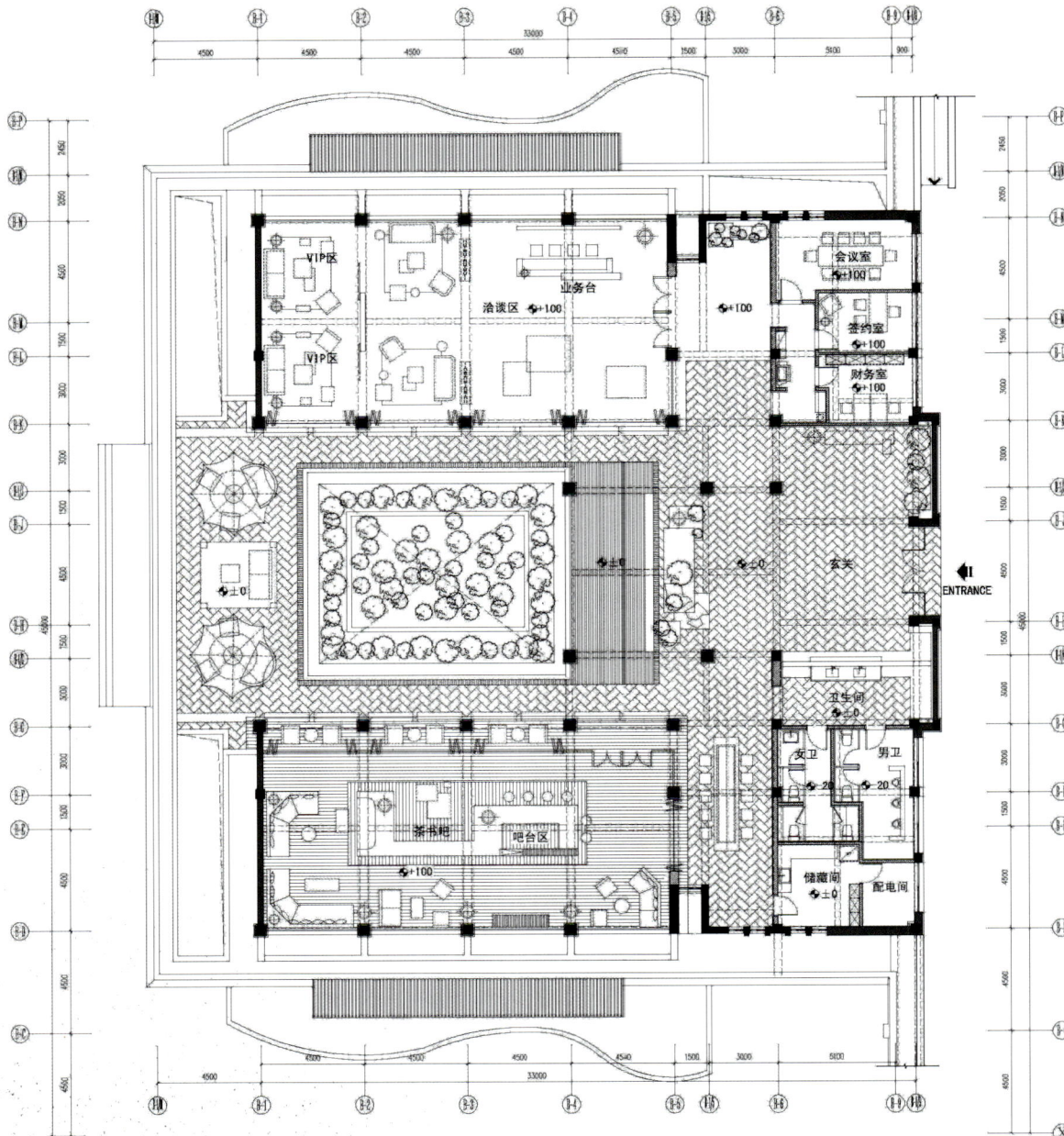

平面图

售楼洽谈区域面积：　240.77平方米
茶室、公共区域面积：　770.52平方米
总面积：　1011.29平方米

The Reception Lounge of Narada & SPA Hotel
——Aesthetic courtyard for literary families!
This space is full of smell of book fragrance and reflects the Hangzhou-style aesthetic principles. In tropical zone close to sea, we try to create an atmosphere that people can relax and ponder calmly and peacefully. And then, this Chinese-style courtyard, which is originally of straight and narrow, now is equipped with a fresh and elegant bar for reading books and sipping tea, a dignified and graceful study room and a reception room with its floor paved with plaster bricks. Here you can read books or recall your memory, you also can stroll idly. Any definition about the space functions is slim.

Bamboo chopsticks, the surprise from touching
The bar for reading books and sipping tea is the most
beautiful scenery of this space.
We expect to endow the surface of the bar counter with more meaning.
Bamboo chopsticks were chosen because bamboo symbolizes gentlemen
with humility and integrity in Chinese traditional culture.
The echoing between the most natural associations about the material
and the space style results in not only a surprise from touching
but also the aftertaste of life and the inspirations from life.

Implanting Simple Texture
Walk through the bar,
you will find that a large area of the floor
is paved with natural plaster bricks without any decoration.
We hope each one who comes here will get natural and true feeling from each step.
This feeling tends to shorten the distances between people and the space or interpersonal ones.

Discovering Life Aesthetics From Details
In the study room,
the 1.80-meter-tall light of primitive simplicity,
the bookcase with strong primitive feeling,
the elegant bird cage at the corner of the wall,
each pot of flower, each tree,
are the full details we tried to deliver the charm and elegance of Hangzhou-style aesthetics.

这是一个充满了书香气息，深谙杭派美学之道的空间。在临海的热带，我们试图营造一个清凉、静谧，可以让人放松和思考的氛围。于是，这个原本中规中矩的中式庭院里，便有了清新淡雅的茶书吧，厚重典雅的书房，铺满灰泥砖的会客区。在这里，可以静静地阅读和回想，可以茶书会友，也可以闲庭信步。任何功能性的定义于这个空间而言都变得微不足道。

竹筷，触得的惊喜！
茶书吧是这个空间最亮的风景
吧台的表面，我们希望赋予它更多的内涵
选择竹筷，因为在传统的中国文化中
竹 是虚怀若谷，高风亮节的君子
对材料最自然的联想和空间的格调彼此的呼应
结果不仅是一份触得的惊喜
更是对生活的回味和启发

植入质朴的纹理！
走过茶书吧
大面积的地板采用自然的灰泥砖铺成，砖上不做任何的修饰
我们希望每个人来到这里，都踩到自然和真实的感觉
这样的感受
往往最容易拉近人与空间，人与人之间的距离

于细节处发现的生活美学
书房1.8米高的古朴长灯
柜中深沉的古意
墙角雅致的鸟笼
每一盆花，每一棵树
我们都试图用更加饱满的细节
诠释杭派美学的风韵和清雅

Pinki（品伊）创意机构

刘卫军

Pinki Interior Design Consultancy Co., Ltd was founded in 2000 by its lead designer and creative director, Mr. Liu WeiJun, which specializes in services of interior designs and furniture designs for Real Estates, hotels, business investors and luxury residences. After nearly 10 years of efforts, Mr. Liu WeiJun and his team have worked together, making Pinki a well known brand in the world of design. It also becomes a member of the CCD, and earns many credits as 'IAID the most influential design studio in China', 'the most worthy cooperating design studio in the world of commercial real estate', and 'the most potential design studio in China'. Mr. Liu WeiJun has also won many personal credits home and abroad which achieves excellence for his company.

　　Pinki（品伊）创意机构，由刘卫军创立于2000年，并担任首席设计师、创意总监。专为房地产开发商、酒店、商业投资商、别墅住宅等提供室内设计及陈设设计服务。经历了近十年的发展，刘卫军与团队的共同努力，使得Pinki成为室内设计界的旗舰品牌，并成为中国建筑装饰协会会员单位，IAID全国最具影响力设计机构、地产界最具合作价值设计机构、全国最具潜力设计机构。刘卫军个人更是获得国内外数百项殊荣，创造了佳绩。

Shiou Shangjiangcheng Chamber

Design Agency: Pinki Interior Design Consultancy Co., Ltd
Location: Fuzhou
Client: Fuzhou Zhengrong Group
Area: 4,660m²

世欧上江城会所

设计公司：Pinki（品伊）创意机构
项目地点：福州
客　　户：福州正荣集团
项目面积：4660m²

'Charm eastern' is the design concept of Shangjiangcheng chamber. In the macroscopical design, construction is well combined with the environment around. Besides, with the cultural history of Gangtou River, the local special cultural custom is perfectly presented out.

Shangjiangcheng chamber is located by the Gangtou River, Fuzhou. The design elements are from Chinese traditional natural cultural life, which well expressed the affinity and visual arts appeal of the on-spot environment.

Series of cultural art space of appeal are created through a common kind of space esthetics as well as ordinary decoration materials. The greatness and purity of this building are also represented in its insides, which realize the design concept. And with the very beauty of Suzhou gardens, the chamber is really of modern orient aroma.

上江城会所主题定位"韵魅东方"，它在整体规划中将建筑与周围的环境融为一体，再结合港头河的历史文脉，展示出独到的文化地域气息。

上江城会所坐落在福州港头河边，以中国自然传统文化生活景象为设计元素，体现现场环境的亲和力和视觉艺术感染力。

通过一种非常熟悉平凡的空间美学和普通的装饰材料，创造系列富有感染力的文化艺术空间，并将建筑的大气、纯净带入室内中，实现了传统与现代结合的设计理念，再配以苏州园林之美感，充分体现出具有现代东方韵味的会所空间。整体设计赋予了上江城会所浓郁的品牌文化个性。

Eric Tai Design Co.,Ltd

戴勇室内设计师事务所

戴 勇

He is a famous influential interior designer among the professionals from Chinese contemporary interior design world. Began his career of interior design in 1992, and started his company Eric Tai Design Co., Ltd. in 2004, Tai's works has won the nomination of Andrew Martin International Interior Design Award which is known as the 'Academy Award' in the world of interior design in 2009.

Eric's design, combining the romantic mood of the East and the elegant luxury from the West, is highly appreciated by his clients and peers for reflecting the noble quality of original elegance.

中国当代室内设计界具有影响力的知名室内设计师之一，1992年开始从事室内设计，于2004年创立戴勇室内设计师事务所，2009年作品入选素有"室内设计奥斯卡"之称的Andrew Martin国际室内设计大奖。

戴勇的设计结合了东方浪漫情怀与西方典雅奢华风格，体现出原创优雅的尊贵气质，备受客户和设计师的高度赞赏。

Reception of Maoshan Mountain East Holiday Inn

Design Agency: Eric Tai Design Co.,Ltd
Location: Jiangsu
Client: Jiangsu A&C Real Estate Development Co., Ltd
Area: 650m²

茅山东部假日精品酒店接待处

设计公司：戴勇室内设计师事务所
项目地点：江苏句容茅山
客　　户：江苏美加房地产开发有限公司
项目面积：650m²

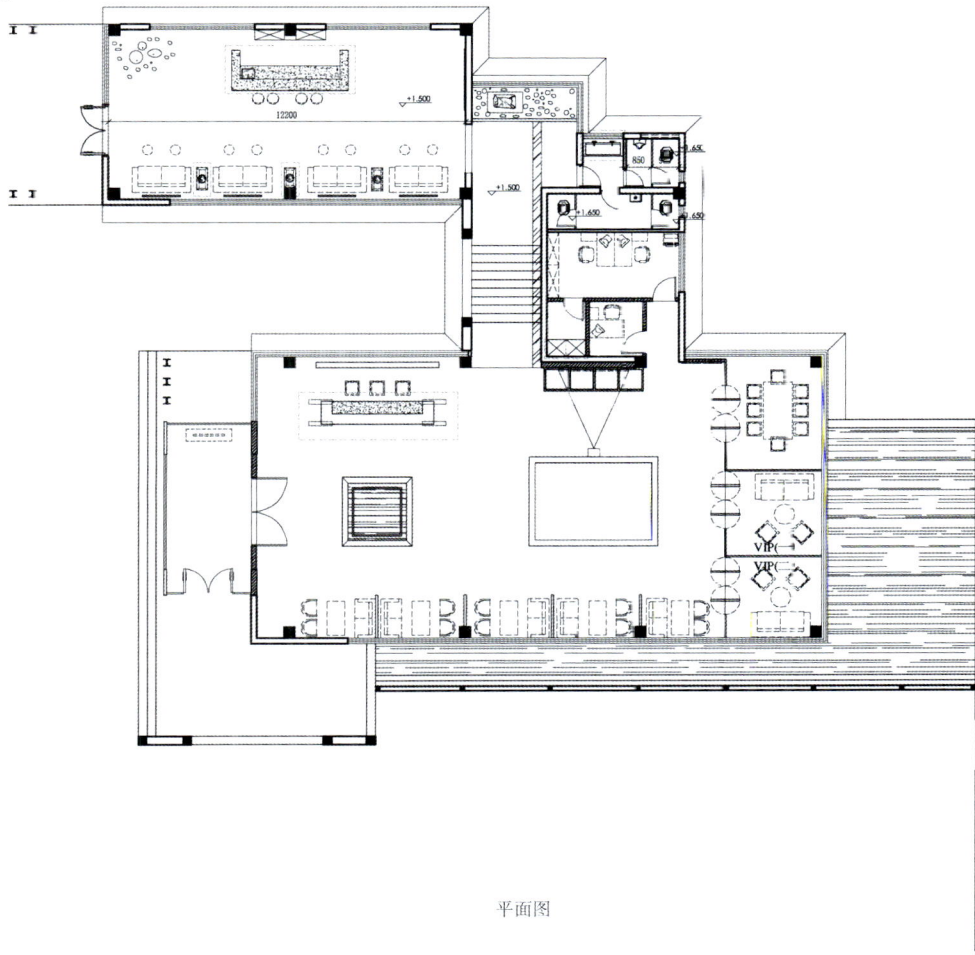

平面图

This project is located in the beautiful landscape of Maoshan beautiful spot of city Jurong. Reaching the summit of Maoshan Mountain, scoping East toward Tai Lake, cloud in the sky and water in the lake appear indistinct. West scope to Chishan Mountain, misty around. Relying on the predominant nature condition, the interspace of architecture intends to lay out the elegant environment inosculated with the culture of Jiangnan.

Walking in it shows the droplight dropped from the ceiling, unite form expanded in alignment, its sculpture brings sharp impact of vision, besides the interspace with unstinted level forms mighty vision focus. Transverse ceiling design is the intention of crossing dropped row with the adorned light. It extrudes the plenty layer feeling, equality lights latented into the rift of ceiling. It let massive top map neat image.

The doorframe just keeps open like that without any decoration, it leaves the transition area to another interior space. At the moment, the flickering wall takes up all idea, it attracts all eyeballs. What makes of that? Take a closer look, it can clearly been seen that dotted deep color stone alternated with green glasses. That's why it gives people the larruping influence scene. In order to match two sets of chairs, candor lines are made briefly, the back side of the chairs carved elaborate snowflake if the white color displays with deep color mat to match the scene. Everytime approaching here, it is nature enjoyment, enjoy the sweetness of art and design.

The color is used green grey as major in the wall of passageway. It assorts with the artistic decoration in Modern Movement, it is riotous in silence meanwhile brings the fashion style. Managing soften colored lines to represent the morbidezza side of the hotel. The brighten color display compared to the whole simplified space, it gives the intensive impact in vision. Couple of benches in red breaked the oppression, increased the layers to expand the space further more.

项目坐落在风景秀丽的句容市茅山风景区，登临茅山之巅，东望太湖，云水苍茫；西观赤山，烟雾缭绕。依托优越的自然条件，在建筑空间内意欲展示融合于江南文化的优雅环境。

入门即见从天花垂列而下的吊灯，统一形状呈队列展开的造型带出强烈的视觉冲击，并且在拔高充裕的立面空间形成了有力的视觉聚焦点。横向的天花格条设计，是刻意交叉于垂列的灯饰，突出丰富的层次感触，灯盏均匀隐于天花隙间，让厚重的吊顶映射出楚楚动人的影像。

石墙上的门洞，留出一个通往另一空间的过渡区，此刻左边吸引眼球的那片闪烁荧光的墙体占据了所有的思想，是什么构造而成的呢？近看时，清晰可见细密的深浅色石材交替着绿色琉璃，才有了这样与众不同的情景感应。为了应景而配的两把坐椅，线条直率简洁，椅背更有着精细的雪片雕刻，若觉白色稍显突兀时以深色的坐垫融合这里的情景。

门后的过渡区，设置一方景观台，以雕塑的形态继续诠释艺术格调，低垂的灯饰以优美的造型成为一体的造景。依级而下，渐入静谧的休息区与洽谈区。以灵活的中轴旋转门作为各功能区间的隔断，各处的环境得以共享。为了形成有效的私密性，旋转门上的玻璃全部贴饰木雕花板，而且这种元素的图案贯穿整个空间。书法墙纸修饰了天花，抬头相望间便是文化气韵在感动心声。

壁廊墙体以青灰色为主调，搭配着具有现代动感的艺术装饰，沉静中带着一点不羁，同时也带出一种前卫的时尚。运用色彩柔化空间的线条，表现出酒店柔美的一面。色彩鲜明的陈设与整个素雅空间形成对比，给视觉以强烈的冲击。几方红凳打破了中间的沉闷，增强了层次感，对空间进行进一步的延伸、扩展。

Chongqing Yongchuan Runjin Garden Sales Centre

Design Agency: Eric Tai Design Co.,Ltd
Location: Chongqing
Client: Chongqing Runjin Real Estate Development Co., Ltd
Area: 950m²
Photography: Jiang Guozeng
Main Materials: rice white Italian travertine, khaki beige marble, black Shanxi marble, wooden sides, solid wood flooring

重庆永川润锦花园售楼处

设计公司：戴勇室内设计师事务所
项目地点：重庆
客　　户：重庆润锦房地产开发有限公司
项目面积：950m²
摄影师：江国增
主要材料：意大利米白洞石、卡其米黄大理石、山西黑大理石、木饰面、实木地板

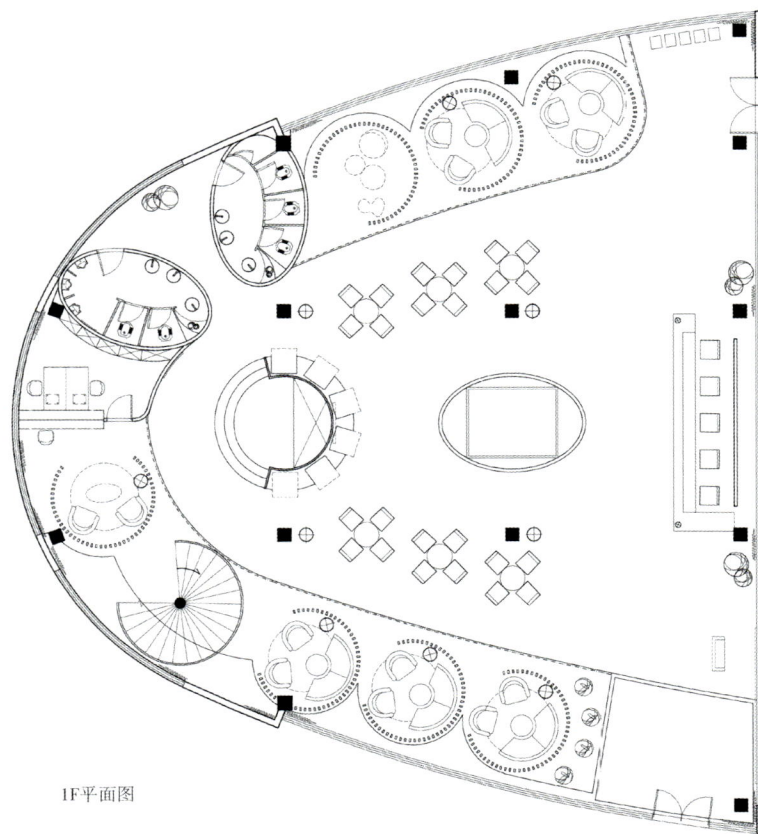

1F平面图

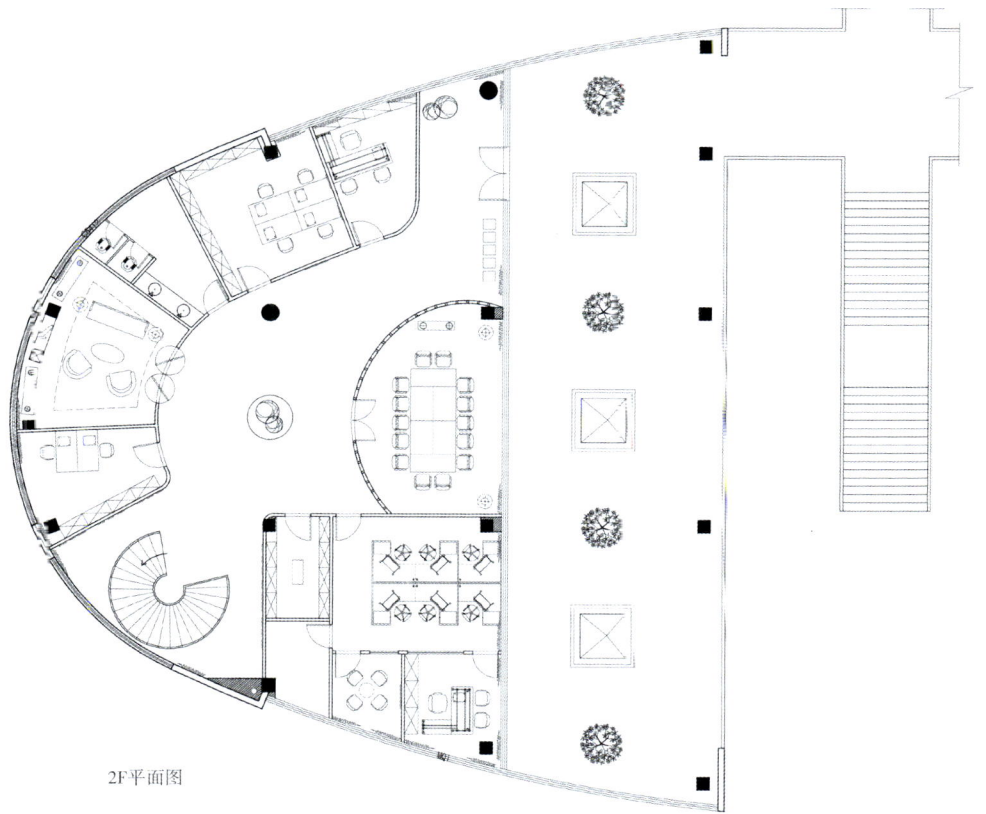

2F平面图

Bamboo represents integrity and condonation. When painting bamboo, the ancients want what they have painted resemble the real object. However, in the design of sales centre of Runjin Garden, the theme of bamboo is incarnated in a modern way.

The design is started with the specific scene of the building and the design inspiration is from half-oval shape plane. Along the sleek outer walls of the building, circular discussion areas are railed in with lathy paling. And the water bar and the platform are also in roundness or oval, together with the spiral staircase to the loft.

Bamboo in mountain gives us a view of exuberances and comeliness. And from it, in design of the exhibition hall on the first floor, we used the image of bamboo. The same material is used. Floors are lain with light travertine and ceilings are decorated with large scale of margins, which set off well the beautiful radians of the dark solid wooden floor and the firm stance of lathy paling. And all these are expressed well in free-flowing lines in the space, the long strip black two-headed lamps on the ceiling, long strip khaki marble in the floor and the cuboid droplight over a model. The very chosen furniture and decorations are contracted and unvarnished with high quality.

The second floor is mainly for reception, conference and office with a style of the exhibition hall on the first floor. The plane is also expressed in a feeling of lines. The semicircular translucent assembly room is enclosed by gray glass and lathy wooden palings, which corresponds well in the design theme. The designers almost combine in a perfect way the flintiness of beelines and the morbidezza of curves.

竹，彰显气节，虽不粗壮，但却正直、外直中通、襟怀若谷。古人画竹，讲究师法自然，极工而后能写意。而在永川润锦花园销售中心的设计中，我们以竹为设计主题，用一种现代的设计语言来表达对竹的情感、用现代的设计手法来传达竹的神韵。

从平面开始，设计师就打破了以往的惯例，从具体的现场建筑特点出发，用一种全新的思维方式来组织平面布局。半椭圆形的建筑平面激发了设计的灵感。顺着原建筑流线型的外墙，设计师用细长的木格栅作隔断，围合出一个又一个圆形的洽谈区。中间的水吧和展台也同样被设计为圆形或是椭圆形，包括入口右手边那通往二楼的圆形旋转楼梯，也以优美流畅的弧度向上盘旋延伸。

那生长在山间的竹林给人呈现的是一片密密的、纤细而清秀的景致。于是，在一层展厅的设计中，设计师借用了竹的意象，采用一种细长的比例关系。深色木饰面的木格栅一圈又一圈，细密的垂直线条给我们带来了竹的韵律。在用材上，我们大面积地使用同一种材料，手法尽量练达。地面满铺着浅色的米白洞石，大面积留白的天花简洁平整，淡雅的背景更加衬托出深色实木地板地台的优美弧度和木格栅笔直挺立的硬朗姿态。天花上嵌着双头射灯的黑色长条造型，弯曲的暗藏暖色灯带，地面上偶尔出现的长条形的卡其米黄大理石，以及挂在模型台上方的长方体吊灯，在整体的空间构图中都转化为线的元素，自然流畅。精挑细选的家具和饰品简约质朴，却又不失档次，同样流露着如竹一般高雅脱俗的品味。

二层主要为接待、会议及办公空间，同样延续了一层展厅的设计风格。在平面上依然是流线型的布局，通透的会议室用灰色玻璃和一根根细长的木饰面格栅来围合成半圆形，呼应着设计的主题。设计师将直线的硬朗和曲线的柔美近乎完美地结合了起来。清风朗朗竹之韵，一枝一叶总关情。古人历来画竹、咏竹，而在这里，设计师重又拾起了这个被人们吟咏了无数遍的题材，用现代抽象的设计语言谱写了一片竹的韵律，静静地回荡在简洁现代的空间中，回荡在重庆永川的御陇山中。

Tashan Club

Design Agency: Eric Tai Design Co., Ltd
Location: Jinan
Area: 500m²

济南他山会所

设计公司：戴勇室内设计师事务所
项目地点：济南
项目面积：500m²

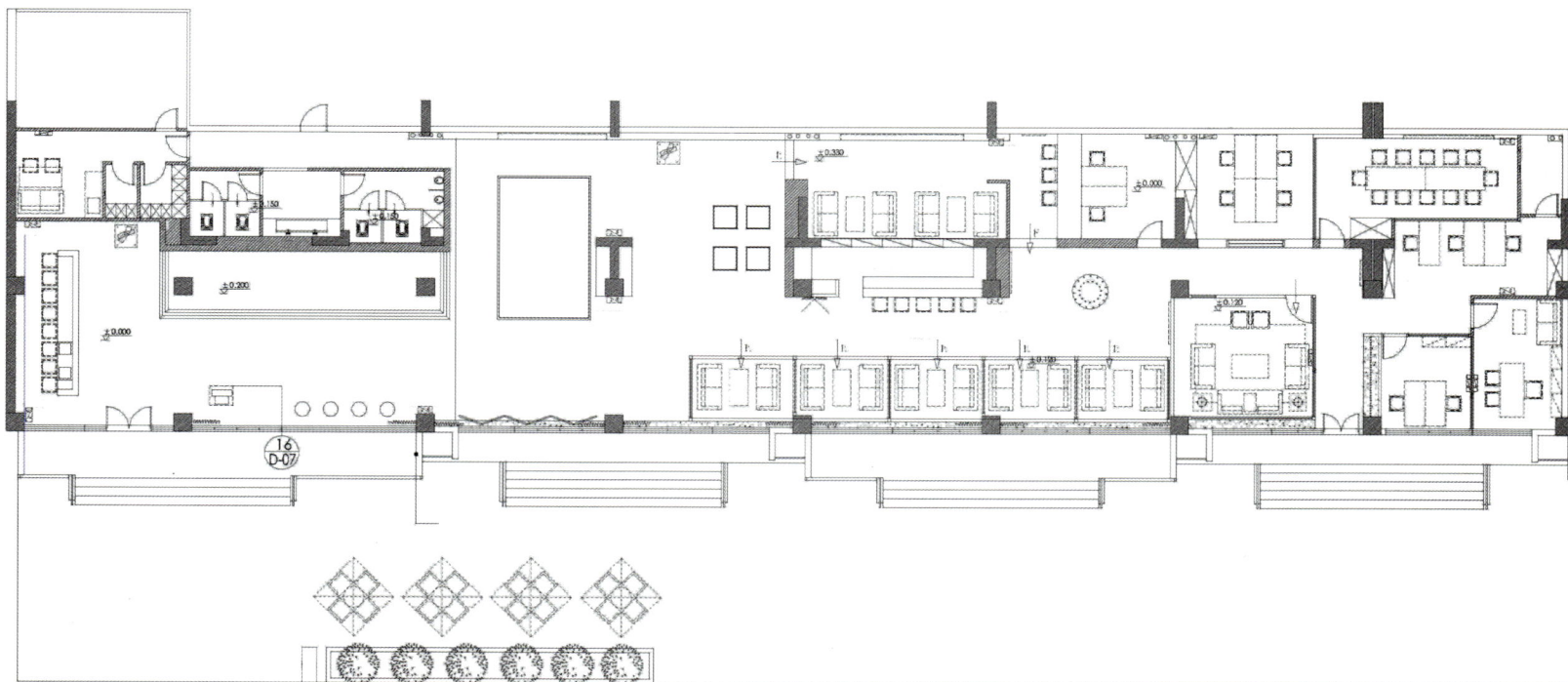

平面图

Tashan is a word of full quaint meanings. The pursuit of its sales centre is to provide a comfortable dainty living environment. The inside of Tashan chamber not only presents the modern new Chinese style but reflects spirits of humanism and arts, in spite of its rigorous modern materials, its classical black-and-white pattern and modern outlook.

When open the door, you can directly see a wall of mountains, flowing lines on the pure white wall. All these help to perform a work of nature, the small pool in front of the theme wall, the green bamboos in the vat and the rippling sound of water.

The strong tall metal columns are just like varied bamboo, contrasting the aroma of success. The very designed paper chandelier is in the theme of mountain stream of ink and wash. In silent nights, it warms every corner in the room and refreshes you.

What designer well designed are not only models but colors. The very added peach painted glass on a desk brings warmth and joy as well as fashion and innervation with pure lights.

　　"他山"深蕴古意之词，此处"他山"销售中心意在追寻人文居住的舒适雅境。虽用酷睿的现代感材质和经典的黑白搭配构筑时尚外观，然而室内延续的不仅是现代新中式的韵味，更融入一份人文情感及艺术表现。

　　开门即见山体造型背景墙，纯美的柔白色调，线条自由休酣，那一座座"小山峰"，在镜面的承接下绵延不断，绕过一道墙，转到另一面墙，似水般柔顺妩媚。主题墙前的小小水池，大缸中的青翠竹筒，轻轻的水声，共同演奏出自然的乐章。刚硬的墙柱与天花倒映出"山峰"的柔美曲线，灵动处自成一幅山水美画。

　　一根根高大坚挺的金属圆柱是变幻后的竹子，带着竹的气息，进一步烘托自然的氛围。设计定做的大型纸吊灯以水墨山涧溪流为题材，黑丝灯罩的外套让它显得幽雅深邃。夜幕中灯光依次照亮，静默的溪流开始缓缓流动，流向每个寂静的角落，让寂寥的夜开始散发生气。

　　光是物体，可以设计为多种表现形态，照射在原本平静的物体上，能勾起深底里的活力，勾勒出唯美的画卷。设计师精心设计的不仅是造型还有色彩，桌面上特别添加的桃红色油漆玻璃，纯净的光线让它倍加姣美妖娆，跳跃在视线里，带来温暖与欢愉，时尚与动感。

Urban Jungle—Earth of Ronghe Club

Design Agency: Eric Tai Design Co., Ltd
Location: Nanning, China
Client: Ronghe Group
Area: 6,500m^2

都市丛林——荣和大地销售会所

设计公司：戴勇室内设计师事务所
项目地点：南宁
客　　户：荣和集团
项目面积：6500m^2

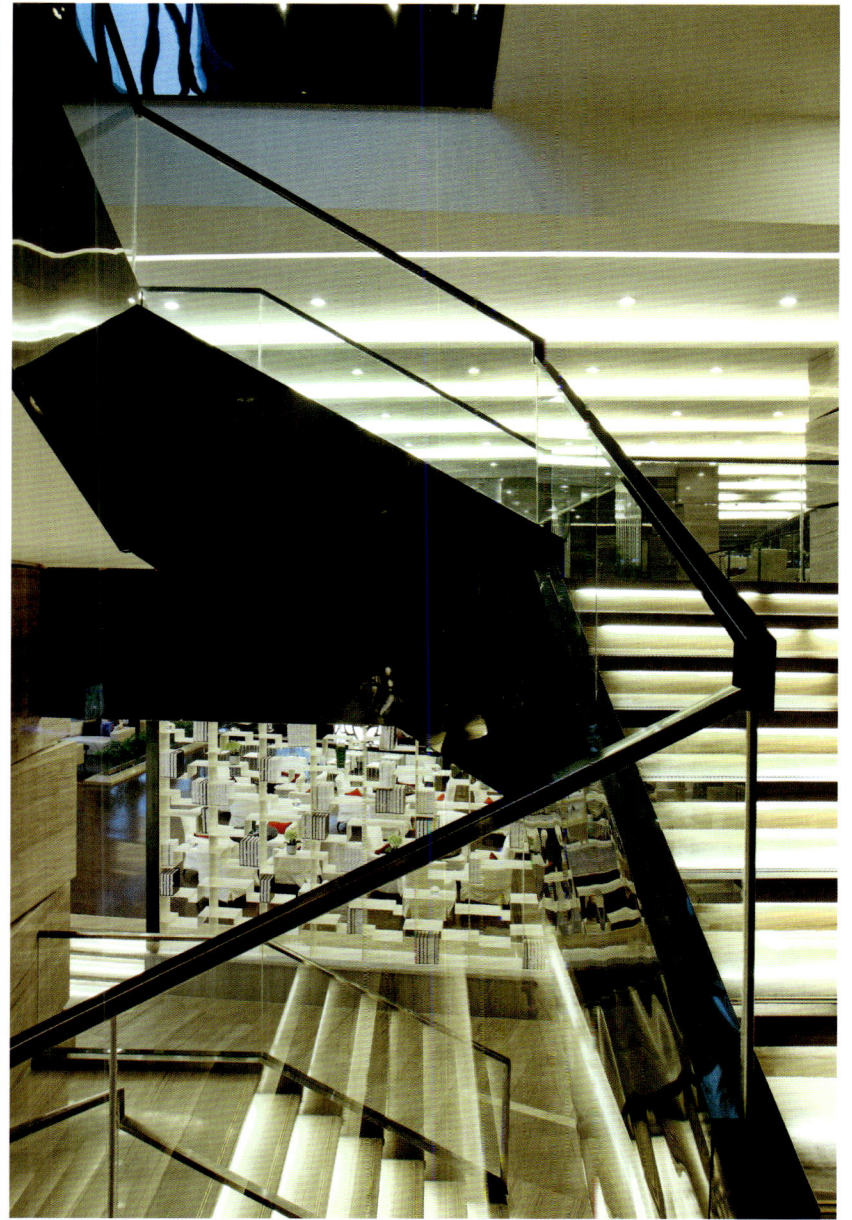

We hope to bring deeper layer subject to the room, although the outstanding & dizzy room is just the vision effect of superficies. For the name of the project, a lot of earth, ocean and woods are brought in as major elements, structural columns are made into Rock effect, dimensions misallocated, and loads of vertical lines are made of arbor in forest, tree trunk. Ceiling is decorated by the stainless steel leaves, there are massive coral shapes above the model tray, weed painting is etched on the glass baluster, and it brings a bit heartthrob to the sharp room.

There is lobby, rest zone, reception, spa & swimming pool facilities on the first floor in the club. Business centre and VIP room are on the second floor. The third floor contains gym and coffee shop. According to the function setting & subject positioning, the first floor is set as the root of tree, with water elements, pool, spa & swimming pool. Second floor is set as trunk of the tree, the corridor is made by vertical style. 'Flower' on the third floor is decorated by flower & grass pattern, such as flower queen wallpaper, mirror plane style etc. Let the tale behind the room.

In the art presentation, if use the color glass to spot the grey room, it will make the room dazzling. The thickly dotted crystal rod of the droplight is like the grown grass in the forest. We can use element to tell a story in room, if you dare to image, then explode your imagination.

空间的酷和炫只是表面的视觉效果，应给空间更深层次的主题。引入大地、海洋及树木的主题元素，把室内的结构柱处理成岩石的效果，大体量的错位；大量竖向的线条是森林里的树木、树干，天花吊落的不锈钢树叶装饰形成绚丽的景象，项目沙盘的上空是大体量的珊瑚造型，玻璃栏杆上蚀刻上水草图案，让一个酷炫的硬朗空间有了点点柔情。

会所一层设有大堂吧、休息区、接待区、SPA及泳池；二层设有VIP包房、商务中心等；三层设有健身中心、咖啡室等。根据功能的设置和主题定位，把一层定为"树"的根，有很多的水、水池、SPA及泳池；二层为"树"的枝干，在走廊上均为竖向的造型；三层为"花"，用花草图案作为装饰，如大花壁纸、镜面图案等手法。让空间造型背后有一些故事。

在艺术陈设上，用彩色的琉璃花在灰空间里作点缀，让空间有一些鲜艳夺目的色彩。吊灯里垂下密密麻麻的水晶棒，像森林里茁壮成长的小草。空间里是可以用元素来讲故事的，只要你有想象，就去大胆地发挥吧！

MoHen Design International

牧桓建筑

Hank M. Chao

Mohen Design International is an award-winning company creating schemes for residential, contract, office and hospitality design in Shanghai, Tokyo and Taiwan. The practice was initially set up by Mr. Hank M. Chao as a platform for cross-disciplinary collaborations. Today the German, Spanish, American, Japanese, Australian and Chinese presses have reviewed the practice's work.

Mohen Design International projects range from public buildings to individual interiors for private clients. The practice has particular experience in the leisure and hospitality industry, developers, focusing on the design of contemporary bars, clubs and restaurants, hotels and private villas. Using a unique language of color, light and geometry, our interiors are sensuous and eventful. Space is carefully choreographed into stylish environments.

　　牧桓建筑 + 灯光设计顾问是一个由建筑师和室内设计师组成的得奖设计顾问公司。由赵牧桓（ Hank M. Chao ）先生成立于1997年，现已逐渐演变成为台北、上海以及东京等地的一个跨国际的设计领域平台。除了建筑、室内设计、家俱、灯光等设计作品外，在专业论述的发表上亦不遗余力。如今，作品已受到包括德国、西班牙、美国、日本、澳大利亚、中国等诸多设计媒体的广泛肯定及刊载。
　　牧桓建筑 + 灯光设计顾问的作品类型涵盖公共建筑物到住家设计。尤其对设计公共商业建筑及与开发商合作有丰富经验，并特别擅长于酒吧、俱乐部、餐厅、旅馆和私人别墅的设计。在对颜色、灯光和几何学的独特掌握和灵活运用下，牧桓设计的空间始终是优美、灵动而别具风格的。

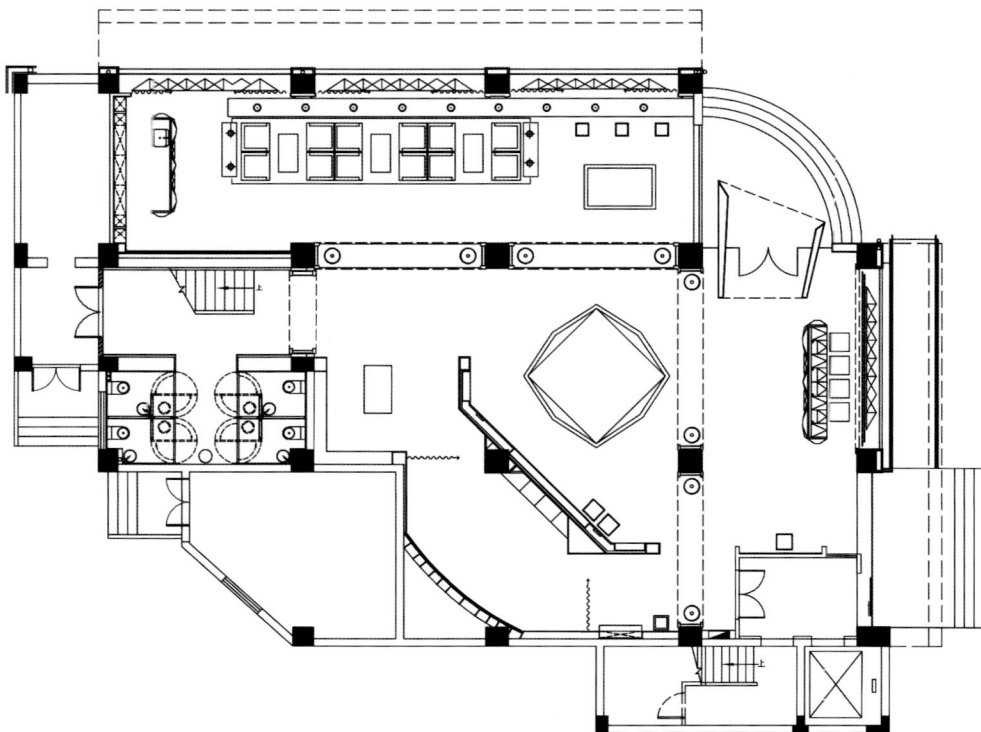

1F平面图

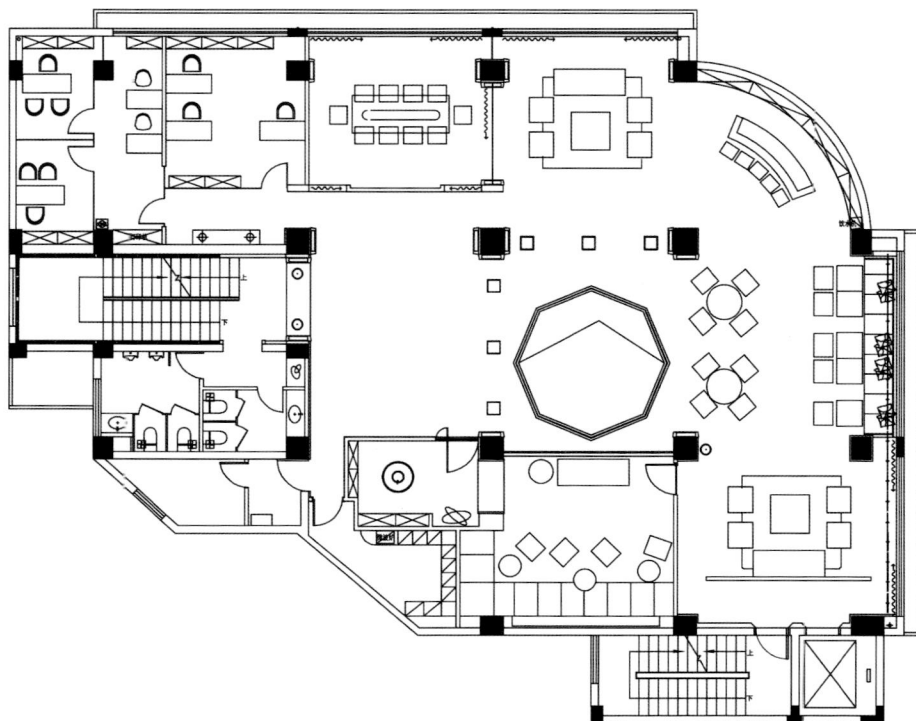

2F平面图

Twinkling Space

Design Agency: MoHen Design International
Location: Shanghai
Client: LVDI
Area: 400m²
Photography: MoHen Design International/
Maoder Chou
Main Materials: marble, custom stainless steel
lighting fixture, acrylic, fiberboard, synthetic
stone

海珀日晖售楼处

设计公司：牧桓建筑+灯光设计顾问
项目地点：上海
客　　户：LVDI
项目面积：400m²
摄 影 师：牧桓建筑+灯光设计顾问、周宇贤
主要材料：大理石、不锈钢定做灯具、亚克力、密
度板、人造石

The base located on the lot of the riverside of Huangpu River, it belongs to the elite Shanghai urban view area. The lot mainly focuses on the usage of future housing exhibition of the entire region. From the perspective of exhibition spacing to initiate the setup of the whole design angle, the designers design to use comparatively glittering and translucent feel to display the harmonize relation and aesthetic feeling of the reflection between the building and the surface of the Huangpu River.

The gate, as the starting point of the space design, likes a tunnel giving visitors an experience of walking along to enter into another space. The designers also pay much attention to the model of exhibition that could form a vertical framework with the second floor. The designers make it possible to see the model clearly from the second floor. Besides, a large stainless steel droplight is specially designed to connect the first and second floor, which underlines the importance of the model as well. The wall behind the model is designed with televisions and buildings sections, which not only helps to introduce latest projects exhibition but marks the future presentation of walls of the building itself. Behind the background wall, there is a projection room for projects introductions detail. The designers use solid acrylic of 4cm in diameter for the ceiling of the discussion area on the first floor. Also the ceiling is specially designed like waves. With the hyperreflexia nature of acrylic, the cubic-effect of the ceiling is well expressed. Sags and crests are well increased on the surface of the reception counter and the bar with the constant repeated use of triangle diamond designs. The washroom is well covered up by the large leather door slices. Specially, you can see carvings twinkle in the washroom when open the door, adding fun of exploring the space. The second floor is mainly for VIP rooms and several thorough discussion areas, which is to appear the space nature of "Twinkling".

Because of it can incarnate the textures; therefore massively utilize diagonal vector and rhombus composition and exercise on high glisten stainless steel. In scme of the lounge area, even use massive transparent acrylic rod in different height hanging on the ceiling forming a matrix pattern. Finally the designers using PAR projection lamp to refract acrylic's nature transparency texture.

Atrium space floor slab in downstairs and upstairs, the designers customize design a highlight stainless steel luminary, using the same shape of module in series connection directly pass the light to second floor, every mono module inside place a single 3W LED light, enhance the refraction texture of the stainless steel, for using as first floor and second floor major visual connection. The floor mainly utilizes white marble in rhombus shape forming geometry shape pattern to reduplicate the other rhombus design language of the entire inner spacing.

本基地位于上海黄浦江边的地块上，属于上海市区的精华景观地带，主要是作为这个区域的未来楼盘展示空间使用。以展示空间的角度出发来设定整体的设计方向，设计师决定以比较晶莹剔透的感觉来显示建筑和黄浦江水面的呼应关系及各种反射、折射的美感。

大门以一种穿越隧道的感觉和形式作为空间的起点。期望来参观者进入此空间宛如进入另一个空间历程。在地块模型的展示上设计师也做了一些巧思，希望与二楼串联成一个垂直的架构关系，除了从二楼也能往下清楚看到模型之外，还特别设计了一款大型不锈钢制吊灯链接上下层关系并更加凸显项目模型的重要性。项目模型后面的墙面则以电视和建筑剖面的手法作为其背景墙，一方面起到介绍项目动态展示作用，另一方面强调建筑物本身日后的表皮语言。在这个背景墙的后面则为投影间，作为深入项目介绍使用。一楼洽谈区的天花以直径4公分的实心亚克力装饰并刻意排列像波浪般的高低起伏，利用亚克力高反射的特性强调其天花立体的形态。接待柜台和吧台则不断重复三角形、菱形的设计语汇增加更多的凹凸面。卫生间则采用巨大的皮革制门片把卫生间掩饰起来，在打开门片后可见到雕刻品闪烁其中，增加了些空间探索上的乐趣。二楼属于贵宾室和几个深入洽谈的实用区块，主要也希望凸显这种"Twinkling"的空间特质。

因此大量地运用斜线向量和菱形构图并运用高反光不锈钢材体现这种质感。在部分休息区域甚至用大量透明亚克力棒悬吊在天花上并排出高低不同的矩阵模式，最后再以聚芳酯投射灯去折射亚克力本身的透性质感。楼下与楼上挑空的楼板区域，定做设计了一个高光不锈钢的灯具，以一个相同形状的模块串联直接通到二层，每一个模块里各放了一颗3W的发光二极管灯，增加不锈钢的折射质感，作为一二层串联的主要视觉连接。地板主要以白色大理石拼出几何造型拼花来重复空间内部的其他菱形设计语言。

HSD水平线空间设计

琚 宾

Designer for Theme Hotels and Restaurants.
Graduated from the School of Design, China Central
Academy of Fine Arts; majored in Interior Design.
Professional interior designer with experiences in
leading scheme and concept design in Hong Kong,
Shenzhen, and Beijing.

Dedicated in studying how Chinese culture can be
applied and innovated with respect to architecture,
involved in various projects by cooperating with
local and international recognized architects and
designers; origins of design: to express individuation
and uniqueness, motives of design: to reinterpret the
elements of Chinese culture.

知名主题性酒店设计专家。
毕业于中央美术学院设计学院，主修室内设计。
毕业后在香港、北京从事专业室内设计工作，主持方
案设计。
致力于研究中国文化在建筑空间里的运用和创新，并
与国内国际多位知名建筑师合作设计项目，以个性化、独
特的视觉语言来表达设计理念，以全新的视觉传达来解读
中国文化元素。

The 5th Sales Centre for Garden City

Design Agency: Horizon Space Design
Location: Shenzhen
Area: 400m^2
Main Materials: crystal pendants, self-leveling cement, white emulsion paint

花园城5期销售中心

设计公司：HSD水平线空间设计
项目地点：深圳
项目面积：400m^2
主要材料：水晶吊线、水泥自流平、白色乳胶漆

The disorder change that comes from light.
The amazing rhythm that comes from rain.
Light and rain are the source of our afflatus.
The simple self-leveling cement floor,
The plain white emulsion paint metope,
And the processes of hanging lines,
It is art,
Performance art,
Installation art.
Which permeated in everywhere of our sales centre.
Everyone,
Everything,
Is part of it.
It is the combination of art and function.
It is the exertion of installation art in interior design.

光给予我们的无序变化，
雨给予我们的韵律之美，
光和雨，
是我们灵感的来源。
简洁的水泥自流平地面，
素雅的白色乳胶漆墙面，
加上一根一根挂线的过程，
是艺术，
是行为艺术，
是装置艺术，
渗透在我们售楼中心的每一个角落。
每个人，
每一物，
都是它的一部分，
是艺术与功能的结合，
是装置艺术在室内空间的运用。

蓝色实业有限公司

刘 鹏

The bluedesign was founded in 1994 and registered into He'nan Blue Industrial Co., Ltd. Designing the office, financial, medical, bar restaurant, estate (sales and sample house) and club is their highlight. They obtained 12 international awards successively, more than 50 domestic awards in recent nearly 20 years of practice. The designer got over 10 honors issued by national authorities and was highly recognized by the society.

Bluedesign confirme that excellent design has a great influence on human. They not only improve the works' design concept and image thinking according to the customer's need, but also take fully the overall projects into consideration and use the design model language to illustrate space. The design language in this established condition to change with creativity is owned by the designers to promote the quality with outstanding planning, creative thinking, detailed analysis and practice. They make promise to service customers and realize the social value of the projects.

蓝色设计(bluedesign)于1994年成立设计室，1998年正式注册为河南蓝色实业有限公司。擅长设计项目有办公、金融、医疗、餐厅酒店、地产（售楼中心、样板房）、会所等。在近二十年的设计实践中，相继获得国际级大奖12项、国内奖项50余项，设计师获得全国权威部门颁发的10多项荣誉称号，得到了业主与社会的高度认可与评价。

蓝色实业坚信，杰出的设计对于人类生活将产生重大影响，将根据客户的需求来完善作品的设计理念与思维创意，更为注重整体项目的综合考虑，利用设计模式语言去诠释空间，创造性地将各种既定条件转化为设计师可利用的设计语言，不断提升设计品质，以优秀的策划、创新的构思、细致的分析及实践承诺为客户服务，实现作品的社会价值。

Kailin International Sales Office

Design Agency: Bluedesign
Location: Zhengzhou Jinshui East Road
Client: He'nan Kailin Real Estate
Limited
Area: 760m^2
Photography: Liu Jiafei

楷林国际售楼处

设计公司：蓝色设计
项目地点：郑州市金水东路
客　　户：河南楷林置业有限公司
面　　积：760m^2
摄 影 师：刘佳飞

To the designers, a good project is undoubtedly a platform to show their talents and release their personalities; To a city, a good construction is not only the regional landmarks in conceptive area, but also a place of intriguing scenery. The sales centre exactly show these ideas, whether from the perspective of real estate developers or from the development mechanism of urban, it is a platform which plays both a display role and an improving role.

As the high grade office building in 'zhengzhou changan avenue', while emphasizing the modern features, KaiLin International sales office as well as pay attention to a character policy: practical, comfortable, beautiful, energy and saving. The twin towers are connected together and look like a tower. Which shows a significant international cooperation spirit, it also combines modern and classical beauty together, and highlights the characteristics the office should have and the characteristics of sedate with enterprising coexist highlighted office should possess. It will be a landmark nature with the modern intelligent office buildings. Because a deep understanding of project, when they accepted clients entrust, they discard the traditional process. They use the more business and international streamline to differentiate space, planning process, spatial functional division for the reception, sand table show, the projection booth, building dish the display. In order to achieve the purpose of display and sale. Designer laid more emphasis on the organization of the interior order, geometrical construction, efficient layout, contrast of texture and the unified color which endowed the sales centre with unique characteristics. The entire space exhibited under certain order and law. Each member here was both interconnected and independence which form a definite gradation not only expressed the strong insistence of the concept of 'stage' but also fully declared the characteristics of stable, solemn, with adequate capital and full of confidence of the building.

对于设计师来说，一个好的项目无疑是施展才华、释放个性的舞台；对于一个城市来说，一个好的建筑不仅是概念领先的区域地标，更是一处耐人寻味的风景，而销售中心就是展现这些理念的载体，无论从房地产开发还是从城市发展肌理的角度来看，它都是一个舞台，一个既有展示作用又有推动作用的城市舞台。

楷林国际作为"郑州长安街"上的高档写字楼，在着力展现现代办公建筑特征的同时，更不忘现代国际流行建筑的八字方针：实用、舒适、美观、节能。大厦将两栋姊妹楼通过中间连接连成一栋塔楼形式，表现了一种明显的国际合作精神，它同时融合了现代与古典的美，突出了写字楼应具有的稳重与进取并存的特点。项目建成后将是一座带有地标性质的现代化智能写字楼。正是由于对项目有着很深的理解，因此在接到业主委托做售楼部设计的时候，摒弃了以往做售楼部常规的流程和手法，以更加商务化和国际化的流线划分空间、规划流程，将空间功能划分为接待、沙盘展示、放映室、楼盘展示，逐步引导参观者去认识建筑，感受建筑，从而达到展示和销售的目的。在设计上更加注重内部秩序的组织和规定、几何的构图、高效的布局、质感的对比和色彩的统一赋予了售楼部独特的个性特征；整个空间展示在一种秩序与规律中有条不紊地进行着，在这里，每一个构件和体块既包容连接又清晰独立，从而形成一个明确的层次，不但表达了"舞台"这一概念的强烈主张和立场，也充分宣示着大厦对外表现稳定、庄重、资金充足、信心百倍的特征。

平面图

金属百叶窗帘（甲供）　　天花遮挡部分　　原有建筑结构　　　　　　成品家具　　　　CT 1 灰色质感地砖　　天花遮挡部分　　原有建筑结构　　　　　金属百叶窗帘（甲供）　　天花看线　　　　CL 5mm厚钢化玻璃植物箱位
原有建筑玻璃幕墙　　　天花看线　　　2mm宽白色玻璃胶填缝　　　　　　　　　CT 2 白色瓷化砖干挂　　2mm宽白色玻璃胶填缝　　原有建筑玻璃幕墙　　通透灯具位　　　BL 1 深棕色皮质硬包
CT 2 白色瓷化砖饰面地面　　成品家具（甲供）　　CT 2 白色瓷化砖干挂　　　　　　　　　红砖砌体地台　　　　　天花看线　　　CT 1 灰色质感地砖　　水吧台位　　　　SS 2 砂光不锈钢饰面踢脚
　　CL-05 墙藏日光灯管(T5)
　　天花看线
　　CT 1 灰色质感地砖

立面图一

SPOKESMAN FOR THE WORLD'S BUSINESS DIALOGUE
对话世界的商务发言人

楷林国味

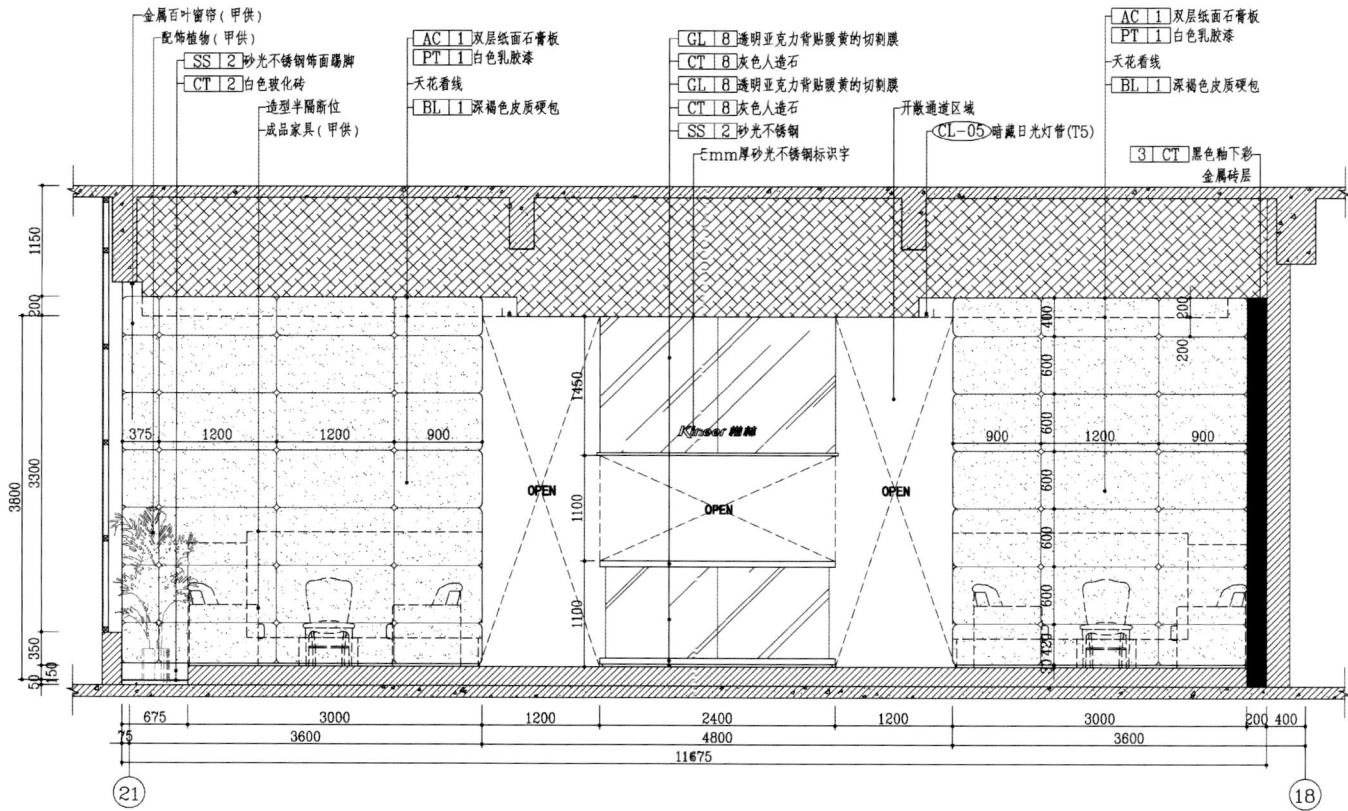

立面图二

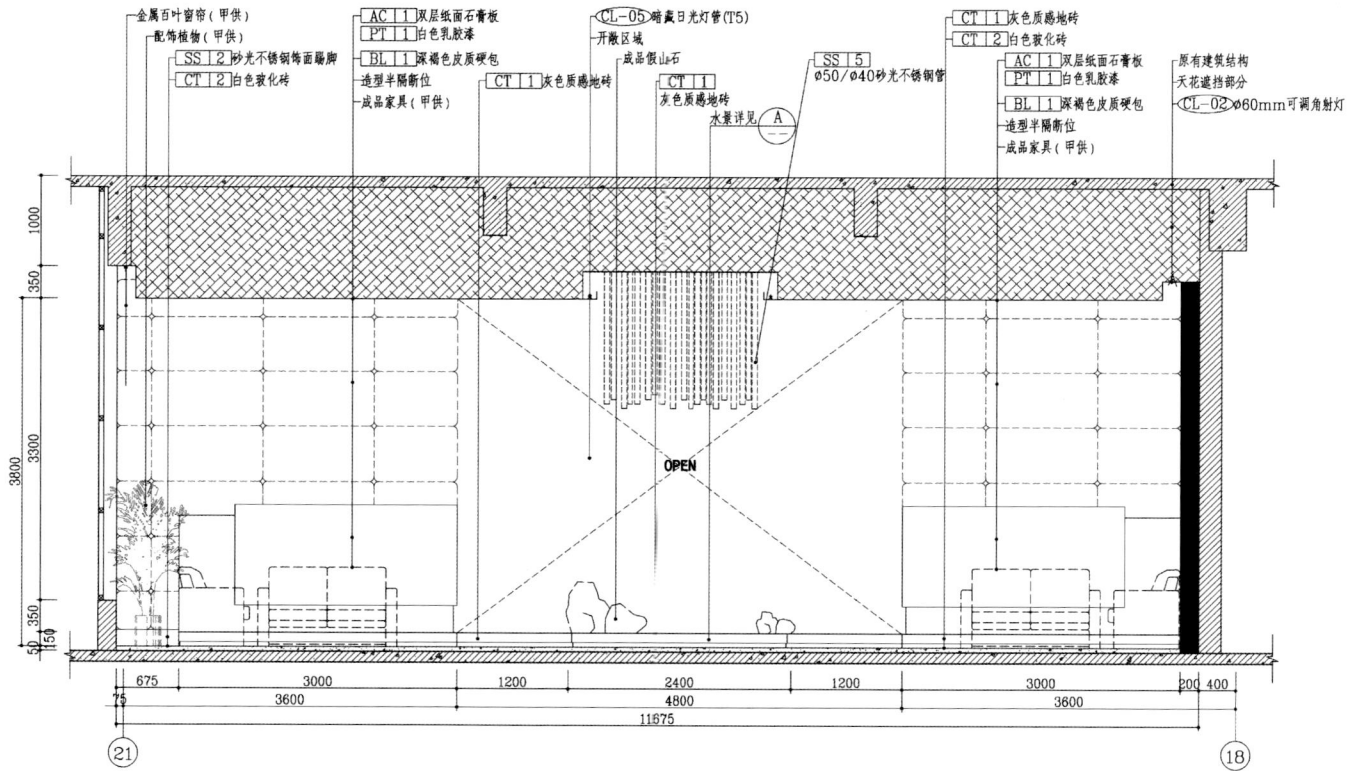

立面图三

立面图四

Heyi Lanze Environment Art Design Co.,Ltd

合宜兰泽环境艺术设计有限公司

邱宜平

Design director of Shenzhen Heyi Lanze Environment Art Design Co., Ltd

The company was founded in the year of 2009. It is a member of Shenzhen Interior Design Association. Ever from the founding, the company's design categories involve hotels, chambers, marketing centres, model houses, restaurants, private villas, etc. The company has established good relationship with many real estate companies. The company offers the most professional design services to clients with different needs with its well educated professional persons and assorted ones.

The HL design team is made up of young energetic creative designers. Their design style is contracted and ingenious. They intentionally combine Chinese traditional cultural philosophy with the west, creating harmony and comfort in varied presentations in use of both traditional and modern design elements. At the same time, a no-time-limit activity space is provided to embody the essence of space.

深圳合宜兰泽环境艺术设计有限公司设计总监。

公司创立于2009年，深圳室内设计协会会员单位，公司成立以来，设计范畴涉及酒店、会所、营销中心、样板房、餐厅、私人别墅等领域的设计，并与很多地产公司建立了良好的合作关系，公司拥有一批高素质专业人才及配套专业人员，在各需求不同的客户中，给予最专业的设计服务。

合宜兰泽设计团队由一批富有激情创意的年轻的设计师组成，风格简约精巧，并着意把中国传统文化哲学融入西方的不同文化，结合古典及现代的设计元素，创出不同的和谐、舒适和不受时限的活动空间，体现空间应有之本质。

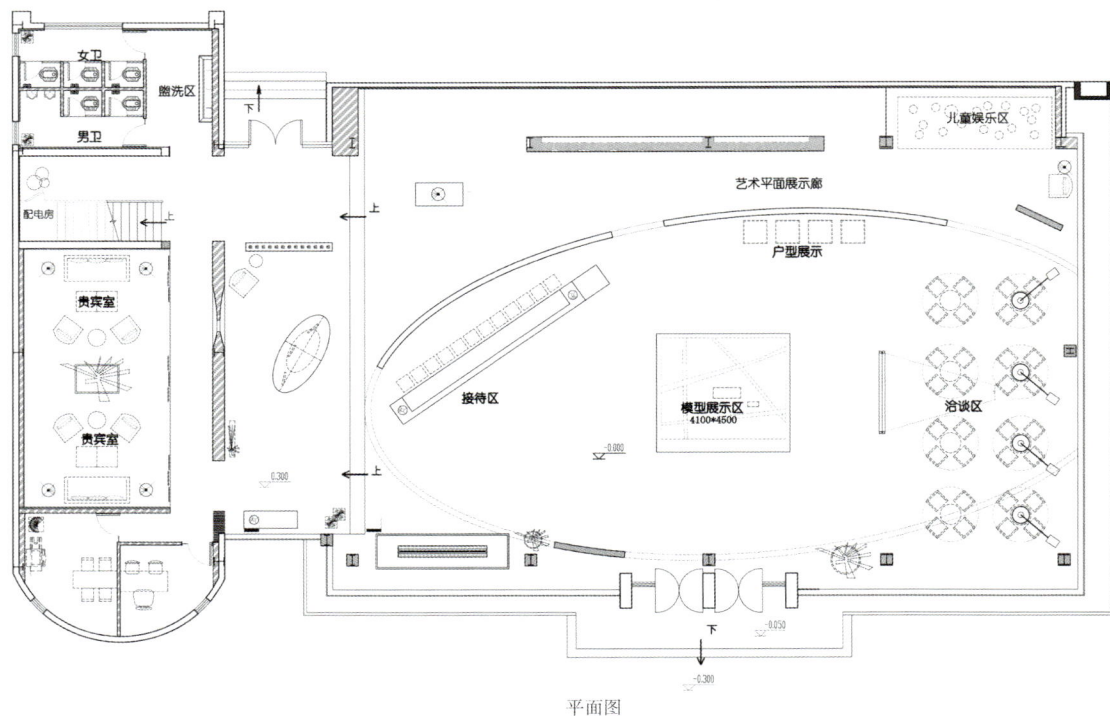

平面图

Xi'an Legend of Star Coin Sales Office

Design Agency: Heyi Lanze Environment Art Design
Co., Ltd
Location: Xi'an
Client: Xi'an Gaoshanliushui Real Estate
Area: 700m²
Photography: Chen si
Main Materials: serpeggiante, oak

西安星币传说售楼处

设计公司：深圳合宜兰泽环境艺术设计有限公司
项目地点：西安
客　　户：西安高山流水地产
项目面积：700m²
摄　影　师：陈思
主要材料：木纹石、橡木

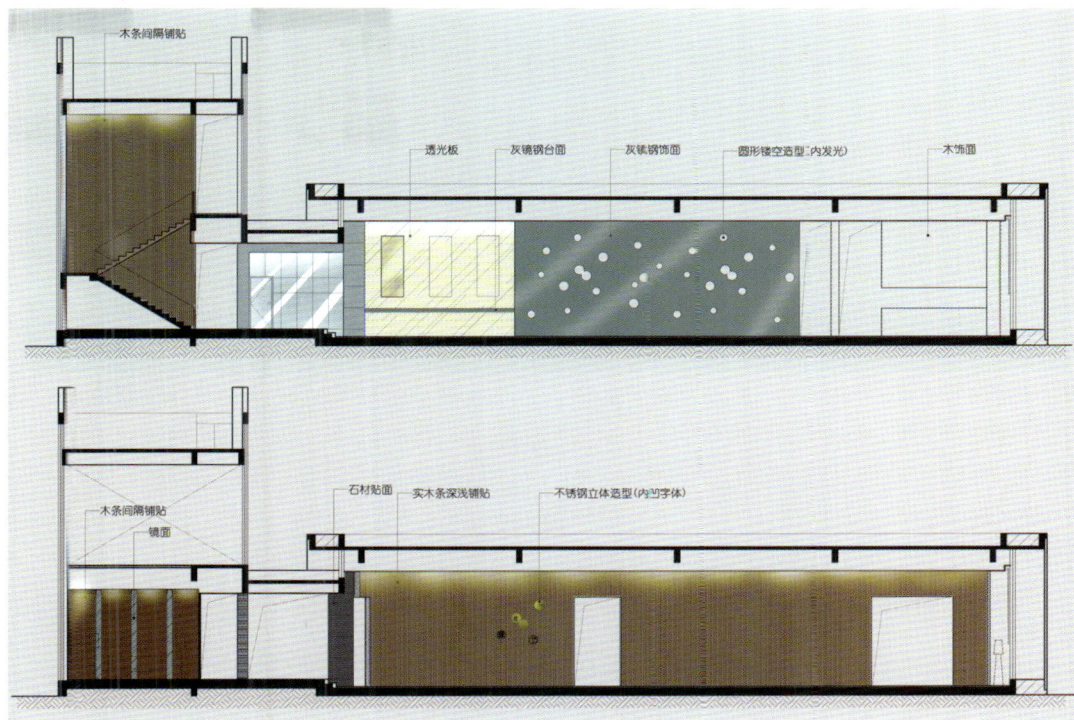

立面图

This project is in Xi'an east second ring. Stars and water are the original design elements. The theme of "legend of star coin" is expressed with space separation and use of light. The insides of the building are in a three-dimensional oval body, which breaks blankness of the inside square space through the use of lines and also enhances arrangements of functions and the look of the project. The effect of stars in sky is perfectly represented in use of wooden lines and acrylic illuminants, which are in full combination of the theme and qualities of the building.

This case has played a key role in transmit the housing culture and its quality in the process of saling.

该项目位于西安市东二环，"星星"、"流水"是本案的设计原始元素，通过空间的分隔及灯光的运用来表现售楼处"星币传说"的主题。室内采用三维椭圆形体，利用线条强化斜面的动性，打破室内单调的方正空间，使其在功能上更有层次，更具观赏性。在材料上大量采用木线条以及亚克力发光体营造满天星的效果，很好地体现出楼盘的主题及气质。

本案在销售过程中对传递楼盘文化以及楼盘品质方面发挥了重要作用，增加了楼盘品质。

都市实践建筑事务所

URBANUS

Under the leadership of partners Liu Xiaodu, Meng Yan and Wang Hui, URBANUS is a think tank providing strategies for urbanism and architecture in the new millennium. URBANUS was set up in 1999 and has Shenzhen and Beijing companies. The name of "URBANUS" derives from the Latin word of "urban", and strongly reflects the office's design approach: reading architectural program from the viewpoint of the urban environment in general, and the ever changing urban situations in specific.

URBANUS has participated in nearly 300 projects of architecture design and urban planning. The completed works range from culture, education, office, commerical space, living, interior, landscape & environment, urban design and research, renovation & regeneration, public art installations for dozens.

URBANUS was selected to the AIA's publication Architectural Record as one of the most influential TOP 10 design institutes in 2005. It won the T+A 2007 China Architecture Deign Institute Award of *Time Architecture*, won the Special Residential Architecture of China Architecture Media Awards in 2008, obtained the China Awards *Architectural Record of Business Week* several times and the WA China Architect Prize of *The World Architecture*, etc.

　　都市实践建筑事务所是由刘晓都、孟岩和王辉主持的建筑创作团体，创建于1999年，目前有深圳公司和北京公司。URBANUS源于拉丁文的"城市"，它表述了事务所的设计主旨在于从广阔的城市视角和特定的城市体验中解读建筑的内涵。

　　都市实践自成立以来参与了近300个重要的建筑项目和规划设计，已建成包括文化、教育、办公、商业、居住、室内、景观与环境、城市设计及研究、改造与更新以及公共艺术装置等作品数十项。

　　都市实践2005年入选美国建筑师协会会刊《建筑实录》年度全球10个最具影响力的先锋设计事务所，获《时代建筑》T+A2007建筑中国"年度建筑设计机构奖"、2008首届中国建筑传媒奖之"居住建筑特别奖"并多次获得美国《商业周刊/建筑实录》中国奖以及中国权威学术奖——《世界建筑》WA建筑奖。

OCT Life Art Place

Design Agency: Urbanus Architecture & Design Inc.
Location: Chengdu
Client: Chengdu OCT Co , Ltd.
Area: 470m²

华侨城售楼处

设计公司：都市实践建筑事务所
项目地点：成都
客　　户：成都天府华侨城实业发展有限公司
项目面积：470m²

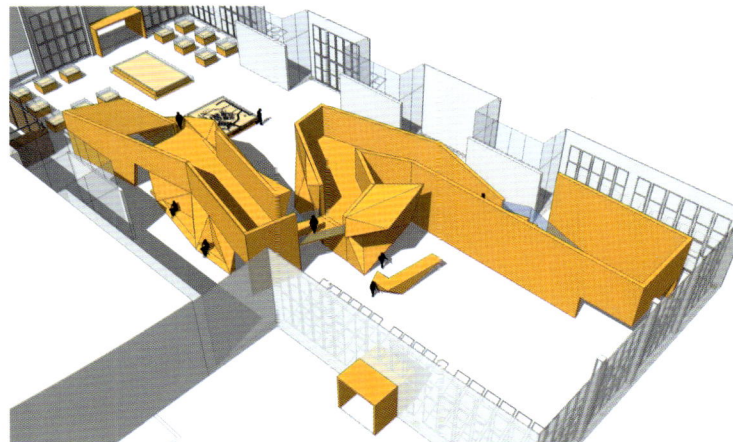

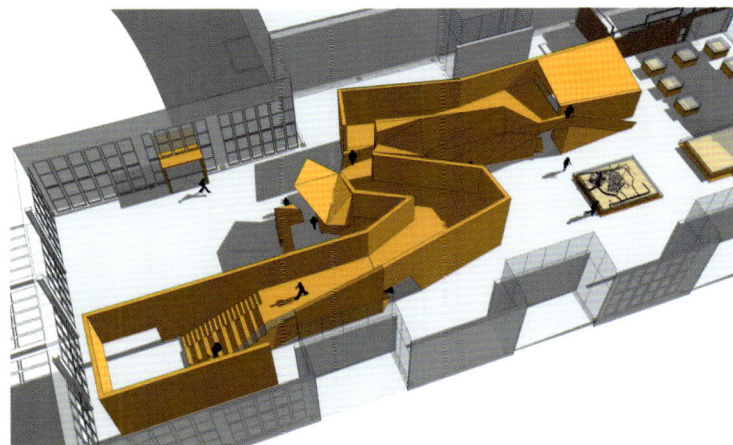

效果图

The relatively small sales office of OCT, a giant real estate development, in Chengdu expresses the theme that art will always integrate with everyday life and architecture, and can even be agreeable with the commercial. In a space originally used as an auto repair shop, the design intent is to keep the building's existing ceiling and walls and create an architectural device independently residing in the building. The new architectural device is sculpted to represent an artificial hill which has reference to the traditional Chinese garden idea of a miniature stone hill landscape where people can leisurely walk around.

This artificial hill is used to separate the loft space into an art exhibition and a sales model exhibition area, and is used to energize the existing loft space by integrating art and functional demands within the commercial and exhibition areas. A main passageway runs on top of the artificial hill forming a meandering gallery space. Sitting spaces blend in and out of the form and a staircase that also functions as seating space during lectures creates an environment allowing for different activities to take place together in one space. The interior of the artificial hill on the ground floor contains conference rooms, offices and a bar. Steel plate is chosen as the main material for the design of the device to amplify the industrial feeling of the original space.

The artificial hill meets the functional demands of the space and injects energy into a once banal place, and creates a space within a space that is not only visually interesting but also invigorating to the senses.

　　房地产公司巨头华侨城集团在成都的一个售楼处的主题为：艺术总是与日常生活和建筑相融，并且能与商业结合在一起。本案原本为一个汽车修理厂，设计意图是保持原有的天花和墙壁，然后创建一个建筑体，独立地坐落在大楼里。新的建筑体被雕塑成一个假山，灵感源自中国传统的园林观念——供人们休闲游走的微型石山景观。
　　假山将阁楼空间分隔成一个艺术展厅和一个售楼展览区，同时也通过将商业和展览区的艺术与功能交融来活跃、调动现有的阁楼空间。假山顶部的主要通道形成了一个蜿蜒的画廊空间。画廊里随处都有座位而且楼梯也可作为座位，这样的设计使得在同一空间里举行多种活动成为可能。底下假山的内部是会议室、办公室和吧台。钢板被选择做主要材料用以增强原空间的企业概念。
　　假山满足了空间的功能需求，并且给最初陈旧平乏的空间注入了活力。多层空间的创建不仅带来了视觉乐趣，也活跃了神志。

成为中国最具创想文化和影响力的企业
The most innovative and powerful
cultural ___ China
切合中国人日益增长的品质需求
cater for the ever-increasing
life quality ___ Chinese
小心台阶
优质生活创想家
An innovator for quality life
borderless imagination
无限想象力
敢为人先地创造
Create the first
为人们提供更具个性化的生活体验
Provide more personalized
life experiences for people

Happiness
City of
Happiness

No.1 Xinghai Bay

Design Agency: Urbanus Architecture & Design Inc.
Location: Dalian
Area: 400m²

星海湾1号

设计公司：都市实践建筑事务所
项目地点：大连
项目面积：400m²

平面图

In the southwest of Xinghai Square, there is a two-story flower colors house standing quietly for years. The building looks tiny among the tall retro style buildings. If sealed its windows and painted it in white, designers could somehow turn it into a white house humourously which would only see in fairy tales.

In order to illustrate this kind of artistic conception, a glass wall is placed in front of the house, making the house gleaming in the reflection of light and environment. There is narrow and long gap between the house and the glass wall. The front yard has wooden floor to raise the ground. Passers-by can also happen to "join in" the party held in the front yard. How surreal the montage feel is!

There is a Chihuly style pendent lamp on the porch. The lamplight flows down to the octagon stone stand which has the sketch of the house and then flows out of the sides. There is no need for 2 floors, so the floor slabs on both wings are hollowed. The white marble is up to 2.4 meters, and above is the ceiling with black mirror. The girders also coated with mirror which makes the ceiling very mysterious. In the unlimited reflections, the pointolites on the black ceiling just like stars on the sky. The colorful glass sticks floating under the "starlight", like fishes swimming freely to the deepest part of the mirror river. Design makes the limited space wider, and changes the ground into underwater.

The decoration involves accessories made of black stainless steel. Origami like toys begin from a flat surface to a three-dimensional funtional bench by cutting and folding such as the wine shelf, bookshelf, candlestick, exhibit stand, etc. have more design feelings.

The flavor of fairy tales makes the entire design unreal and mysterious. This is

exactly what its function is, because the function of the sales centre is to tell an attractive fairy tale.

在大连星海广场西南的一大片空地上，几年来一直默默伫立着一幢两层花色的小楼。在该地段周边林立的复古风格高层中，这个小楼微小的体量倒是有些轻松。如果封上它的窗户，涂成白色，则可以很幽默地把它还原成纯洁的童话。

为了突出这种虚幻的意境，白楼前面挡上了一层玻璃墙，使白楼罩在反射的天光和周边环境中，若隐若无。玻璃墙与楼体间狭长的户外空间，用室外木地板铺成抬高的前院。街上匆匆而过的行人，会忽然间和院子上悠闲的派对叠合，这种蒙太奇有些超现实。

入口门厅挂着一盏奇胡利式的吊灯。灯具象从水面涌向水底的一群生命体，降到刻着楼盘区位图的八边形黑色石台上，再顺着光芒向两边的区域溢出。由于并不需要二层面积，室内两翼的楼板被挑空。高度2.4米以下的是从地面翻上来的白洞石，其上是用黑镜相互反射出的无限空间。结构梁也做了镜面处理，使人一时无法辨别顶上发生了什么。在无限的反射中，黑色吊顶上的点光源形成了满天繁星。星光下浮游着大的彩色玻璃棒，像鱼群一样自由自在地游向镜面最深处，把有限空间变得更广阔，把地面变成水底。

设计味稍浓一些的还有黑钢板做的装饰元素，像折纸般的玩具，从一张平整的平面开始，通过切割和折叠，变成立体的功能性台架：酒架、书架、烛台架、展架等等。

童话的韵味使整个设计罩上一层超现实的色彩。这倒符合功能，因为这个营销平台就是在讲一个让人着迷的童话。

Guangzhou Fangwei Decoration Co.,Ltd

广州方纬装饰有限公司

邹志雄

Guangzhou Fangwei Decoration Co., Ltd was founded in 1987 in Guangdong Zhuhai special economic zones by the chief designer Zou Zhixiong. It is an architectural decoration enterprise B grade of both design and construction approved by the China Association of the Interior Decoration. Also, it is the governing unit of the Guangdong Decoration Association of the Industry, a member unit of Guangzhou Building Decorates Association, a member unit of the Chinese Interior Decoration Association Membership and the governing unit of the IDA International Designers Association, a member unit of Chamber of the Upholstery of National Association of Industry and Commerce.

For years the company has consistently advocated low carbon, carbon emission reduction and environmental protection. The company pursuits for the most excellent design concept as well as the strictest quality standards, and takes quality as the life of enterprise. She is wholeheartedly for the customers. And she strives for the trust, respect and satisfaction of clients. To accelerate the standardized, specialized enterprises development, and they strive to establish the first-class internationalized, collectivized, society respected decoration brand.

　　广州方纬装饰有限公司由方纬总设计师邹志雄先生1987年始创于广东珠海经济特区，是经中国室内装饰协会核定的建筑装饰装修工程设计与施工双乙级企业，是广东省装饰行业协会理事单位、广州市建筑装饰协会会员单位、中国室内装饰协会团体会员单位、IDA国际设计师协会理事单位、全国工商联家具装饰业商会会员单位。

　　多年以来公司一贯提倡低碳、减排、绿色的环保设计理念，追求最出色的（创意）设计，追求严格的质量标准，视质量为企业的生命。全心全意服务于客户，努力追求客户的信任、尊重和满意。加速正规化、专业化企业建设，努力将公司建成一流的国际化、集团化、受社会尊重的装饰品牌。

Guilin Bowangyuan Sales Office

Design Agency: Guangzhou Fangwei Decoration Co., Ltd
Location: Guilin
Area: 850m²
Photography: Chen Si
Main Materials: black Mongolian lamed slab, rust steel plate, steel bead curtain, etc.

桂林博望园售楼部

设计公司：广州方纬装饰有限公司
项目地点：桂林
项目面积：850m²
摄影师：陈思
主要材料：蒙古黑烧板、锈钢板、钢珠帘等

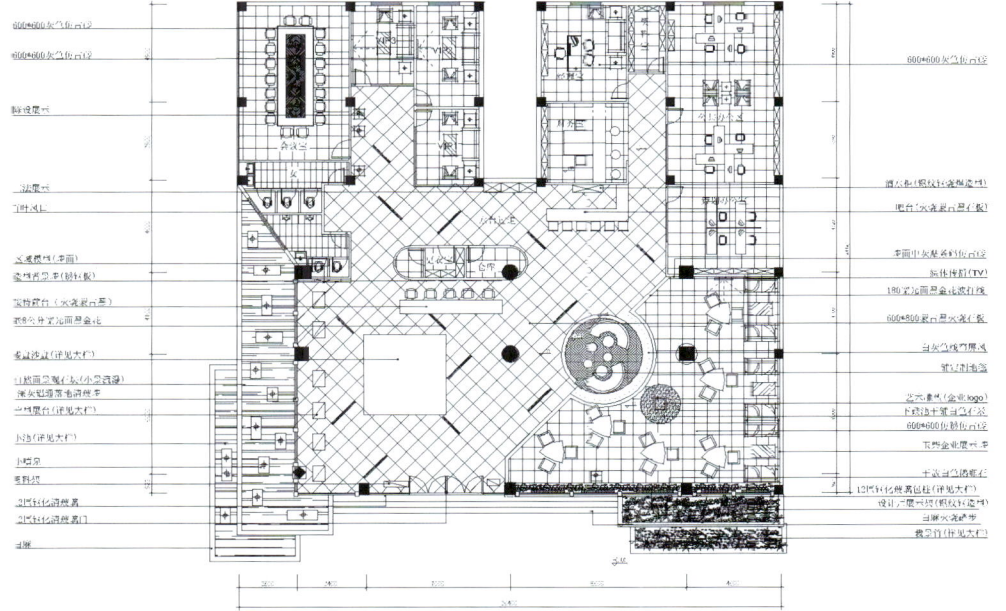

平面图

In the design process, the designers have fully considered the various personalities and characters of each functional area. Areas are well separated in the use of the diversification of structures and shapes and the arrangements of lights, which is of high functionality and practicality. As a whole, dealing with details of lights and changes of space is programmed taking roles of transition while in details, ultra modern art style and rich imagination and creativity are adopted, which combine creativity and conciseness, leading people to feel the space with the integration of funtional taste and concise fashion.

在整个设计过程中，设计师充分考虑到各个功能区的个性和特点，利用结构造型变化和灯光的调整，将各个区域有机划分开来，体现极高的功能性和实用性。在整体上，设计师将灯光和空间微变的细部处理进行规划，起到有效的过渡作用。在细节处理上，又大胆采用超现代的艺术表现手法和极具想象力的构思，将创意和简约因素融入其中，使人感受到集功能品味、简洁时尚为一体的元素空间。

Zhaoqing Hongjingjinyuan Sales Office

Design Agency: Guagnzhou Fangwei Decoration Co., Ltd
Location: Zhaoqing
Area: 300m²
Photography: Chen Si
Main Materials: beige royal marble, light emperador, dark emperador, rustic tile, wallpaper

肇庆鸿景锦园售楼部

设计公司：广州方纬装饰有限公司
项目地点：肇庆市
项目面积：300m²
摄 影 师：陈思
材　　料：皇家米黄大理石、浅啡网、深啡网、仿古砖、墙纸

Elegant European style is the theme and primary hue of the case. The classical outside together with the carefully arranged inside combine cultural tastes and exotic cultures well, which make the sales office filled with international, fashion and culture atmosphere and full of dreams. Windows here and there ensure brightness and throughflow and the soft light of crystal lamp provides an easy comfortable environment. The great sand table of the housing is just by the left of the door, adjacent to the resting area, which is convenient for clients to see the minimized version of the housing.

典雅的欧式风格是本案的基色调和主旋律。经典的外部建筑加上内部的精心布置，将文化品位与异域风情巧妙融合，使售楼部洋溢着国际时尚文化气氛，几乎让人产生梦幻般的感觉。整个售楼大厅因布满了窗户而显得明亮通透，加上水晶灯柔和的光线的点缀，又不失温馨之感，更给人以舒适、安逸的感受。巨大的楼市沙盘置于进门左侧，与休息区相邻，便于客户在休息之时也能欣赏到整个楼盘的缩略版。

上海凯艳装饰设计有限公司

张质生

Cayenne/Mulan Interior Design believes that excellent works are made due to continuous digging, attrition and resonation. After 25 years' of efforts, Cayenne bring in Taiwan the highly valued idea of advanced home decorating, interpreting international bearings of Taiwan's architectures, and a favorable reception from the design industry. Moreover, through cooperating with the industrial workers, research and development departments, Cayenne separated the sectors of management and control, built up a resource integration system for home designing, and made itself one of the most competitive brands in the designing industry in Taiwan.

Cayenne refuses to follow an inflexible pattern; it tends to communicate and listen, which leads to combining of the users' needs and the design itself. With an emotional connection between spatial designs and the users, Cayenne interprets the infinite aesthetic tastes in order to construct the personality of harmonious space. 'Diligence and Trust' is the quality of Cayenne, and to create 'Comfort and perfect living space' is its pursuit. Gathering power from elites and integrating resources, Cayenne is paying all its effort to provide perfect house designs and exquisite lives for its customers.

　　凯艳认为，优秀的作品只会产生于不断的发掘、磨合和共鸣之中。二十五年心血之作，凯艳将国际名仕所推崇的先进家居理念深入台湾，将台湾许多著名建筑的国际化气质演绎到极致，深得业界的好评。更通过产业工人、工艺研发、管控分离等诸多环节的紧密配合，在相关产业领域进行深度家居集成和资源优化整合，从而跻身台湾最具竞争力的品牌梯队之一。

　　凯艳室内设计不拘泥于特定风格，通过充分的沟通与倾听，将使用者的需求融入设计思想中，凝聚出空间与使用者之间的感情，并辅以无限美好的内敛美学品味，建构出和谐的空间个性。"勤奋、致信"是凯艳木兰的品质，"开拓、进取"是凯艳的精神，创造"舒适、完美的生存空间"是凯艳的追求，不断聚合精英力量，整合行业资源，为服务客户创建完美家居与精致生活而努力。

Reception Centre for Giant Egg, Taiwan China

Design Agency: Cayenne/ Mulan Interior Design
Location: Taiwan China
Client: High-Yes Corp.
Area: 7,350m²
Photography: Cheng Xiangyi

新巨蛋接待中心

设计公司：上海凯艳装饰设计有限公司
项目地点：中国台湾
客　　户：台湾海悦广告有限公司
项目面积：7350m²
摄影师：程相怡

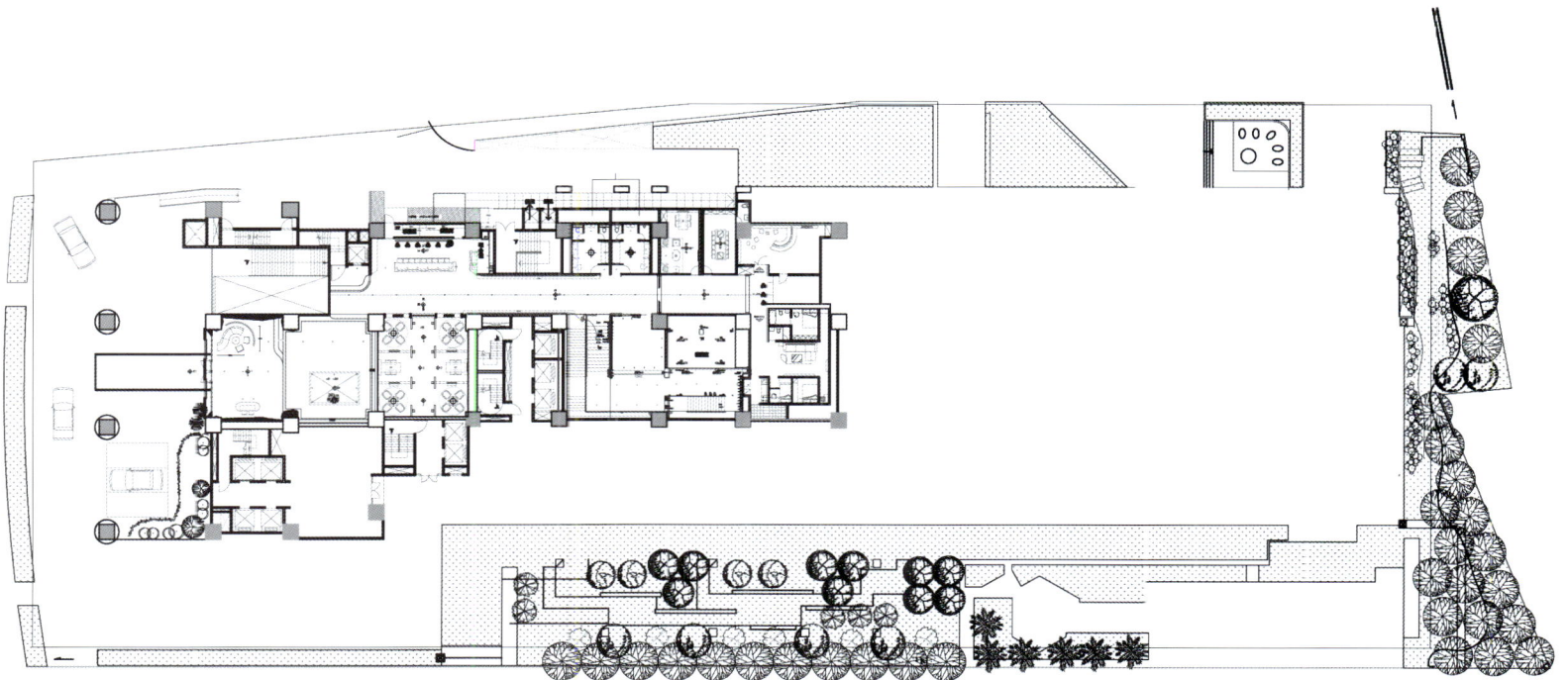

平面图

Giant Egg has made a great success once it is open to public sale. This astonished the whole architectural industry. The project is invested by San Yuan Institution and Fu Qian Industries. It is located around the exit of No.4 MRT, occupying an area of 8,369m^2.

The outlook of this reception centre takes 'new era, Giant Egg' as its design policy.

Its interior design is also a selling point, which reveals the designer's cultural taste. Its furniture layout and color shows the designer's characteristics and aesthetics. From the entrance, one can sense its special concept. From the light and dark color of the chandeliers, one gets closer to the nature. The corridor and water view is segmented by wood fences, which not only keeps its continuity, but also provides a feeling of standing between mist and clouds. Without extra styling, its key part makes it a fabulous design, concise as a whole, generous and smooth. Its dark color gives it an elegant style. The inside water view, flowers and plants create a unique and refreshing view.

The interior design matches its exterior, with parquet floor coordinating with the ceiling. At the same time, it segments functional zones using harmonized soft curves to achieve unity. Using expensive and high quality materials, the project is a symbol of luxury and glory. It matches with a saying that, 'People create surroundings, at the same time, surroundings also shape people'. What this project reveals is the sense of cultural charm and the original intelligence of the designer. To let residence feel the warmness and relaxation is also an aim of this project. By proficiently using designing techniques in matters of light condition, heat condition and air quality, the designer successfully present a leisure culture of simpleness, nature and exquisiteness.

　　"新巨蛋"甫一公开即造成热烈抢购，轰动整个建筑业界！它是由三圆机构及福纤实业投资兴建，正新埔捷运站四号出口，基地壮阔约8369m^2。

　　接待中心外观以"新纪元，巨蛋"作为整体设计方针。

　　本案室内的设计也是一大亮点，设计是文化品味的体现，在无言中传递着特有的文化气息。室内的空间布局、家具的陈设、摆放、颜色都体现出其个性特点，体现文化内涵，体现审美情趣。本案一进门便引人入胜，深浅对比，光影斑驳，其透明映射的吊灯使人亲近了许多，走道与水景，用木格栅虚隔，保持良好的连贯性。若隐若现的感觉！简而没有多余的造型，却在关键部位渲出精彩，简洁而整体，大气而流畅。深色调彰显高雅格调。水景，植物花草，自然而然，乱中有序，加注了空间的灵活性，自由营造不同的装饰效果。

　　室内吊顶造型和建筑的外观造型一致，地面的拼花一方面是和吊顶相呼应，另一方面将功能区域分割明显，用协调、柔美的弧线，达到协调统一。采用高档、新颖的材质，体现一种尊贵，华丽！有一句话"人创造环境，环境也会影响人"，这话很有道理。本案的设计，展现出一种文化的韵味，一种时尚的味道！处处透露出设计师的心思细腻、独具匠心！每个人置身其中都会从内心散发出一种微笑，一种惬意感由心而生！让人感受到温馨、放松。这也是设计的目的之一。一个高舒适度的环境，不仅注重内部、平面空间关系的组合，硬件设施的完善，更注重光环境、热环境、空气质量等等这些综合因素。这才是本设计的重点。在这个基础上体现出较高的文化品味，装修、设计达到舒适、生态、文化、艺术理念，表现了质朴、自然、精致的休闲文化。

Reception Centre for Beitou Tomita, Taiwan China

Design Agency: Cayenne / Mulan Interior Design
Location: Taiwan China
Client: High-Yes Corp.
Area: 1,600m²
Photography: Cheng Xiangyi

北投富田

设计公司：上海凯艳装饰设计有限公司
项目地点：中国台湾
客　　户：台湾海悦广告有限公司
项目面积：1600m²
摄 影 师：程相怡

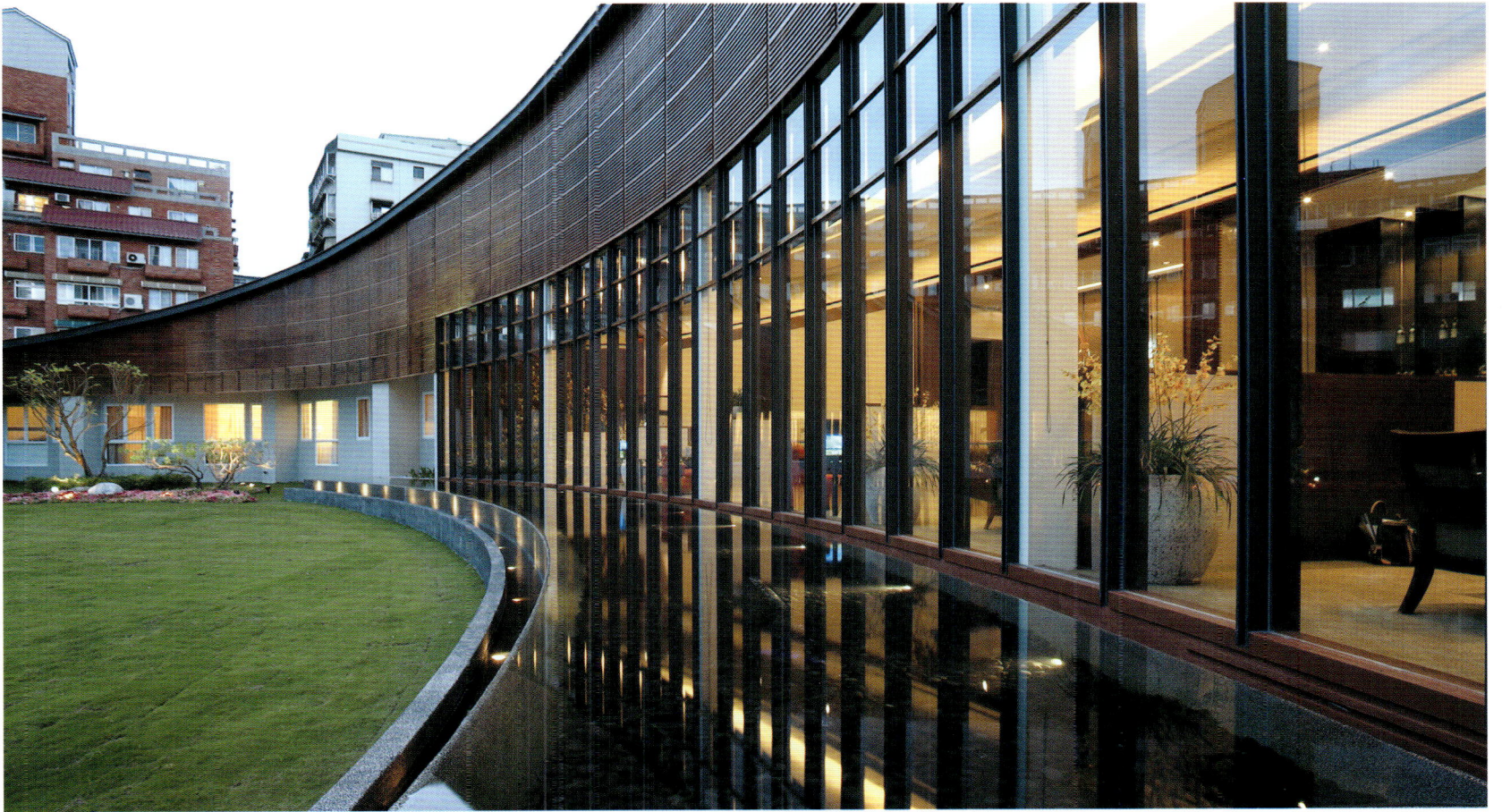

In the design of Reception Centre For Beitou Tomita, its exquisite visual effects and overall steady color tone, makes it a project with luxury, which also conveys creativeness in its steadiness. It reveals the concept of fashion. The image of the entrance is like a fish tail, indicating that the reception centre will be operating smoothly as fish swimming in water. Its design also uses soft curves.

The plan of this project is simple, lively and natural, with a complete generous floor plan. It conveys the feeling of nobility and at the same time denies fake luxury, making itself the architecture with truely high quality. The exterior design of this project mainly uses wood as its material. Many designers have a passion for wood. The love for natural timber is no lesser than a bridge designer's preference to cable-stayed bridges. The completed functions of the project is to reach the standard of 'having everything in it', especially the business functions as meeting customers and having leisure.

This project shows the wholeness of space and unity of style. According to different functions of different rooms, the designer makes out plans for advocating natural simplicity and rational rules, in order to create a project with well-distributed proportion, creative form, well-matched materials, and efficient closure that is convenient for maintenance. Every part of its structure is tightly organized, with spaces intertwining in order, obviously maintaining every element on each constituent. Through changing spatial images between transiency and actuality, it creates harmony between separate parts and the whole space, especially emphasizing on its completeness, dignity and elegance. This design avoids narrowness, constriction, complexity, cliché and stiffness. It emphasizes on playfulness of space. The designer strictly controlled every material in his color-matching, making it a reasonable closure.

The interior design of the project is simple with wholeness in its style. In every part, the designer adds accessories of sculptures, mural paintings, green vines, and wrought copper ornaments. In speaking of detailed decorations, the designer has taken into account the elements of the sky, earth and the wall, making it a perfect example for decency and elegance, fully presenting the taste of the reception centre. Entering the centre through the lobby, one sees dark-colored wood floor, well connecting the space between inside and outside. Also the public spaces between outdoor garden, fountain, plants, and each room have been specially handled, as well as the entrance platform and the corridor. All the grey areas are fully utilized to provide a superior connecting point. From visual feelings to inside comprehension, the reception centre provides with it an extravagant sense. The architecture and surroundings forms beautiful scenery. The whole space are active or passionate, which fully elevates the quality of the space, and expresses the living environment and living method.

北投富田接待中心的设计中，精简曲线的视觉效果，配合高级材质，整体运用了稳重的色调，蕴含着淡淡的奢华感，整体设计沉稳中带着新颖，高贵中透露着时尚的设计概念。入门外观像鱼尾，寓意着如鱼得水，整体采用柔美的曲线。

设计方案在设计方面简洁、明快、大方，平面布局完整、大气，富有贵族气质，杜绝虚假的奢华，体现真正的"高品质建筑"。功能与风格需紧密结合。室内外多采用木材，木材是众多设计师的情结，对于天然木材的偏爱，毫不逊色于桥梁设计师对钢索斜拉桥的偏爱。完善功能应做到"应有尽有"的原则，应具备一定的商务功能（会客、休闲）。

本案体现了空间的整体性，风格的统一性。设计师根据不同功能对相应的房间立面作出处理方案，提倡自然简洁和理性的规则，比例均匀、形式新颖、材料搭配合理、收口方式干净利落、维护方便。整个内部结构严密紧凑，空间穿插有序，围护体各界面要素的虚实构成比较明显，通过虚实互换的空间形象，取得局部与整个空间的和谐，强调空间的完整性和高贵、典雅感。避免空间的狭窄、压抑感、繁杂、老套、呆板的造型，强调空间的趣味性。严格控制每种材质的色彩搭配，合理收口。

室内陈设单纯，符合风格的整体性，局部可考虑雕塑、壁画、蔓藤植物和线形图案或锻铜饰件。细部装饰结合整体（天、地、墙面）考虑，不失庄重和优雅。将接待中心的大气与品味表露无遗。沿着接待大厅往里，铺着暗色木地板的大厅中央，注重室内和室外空间的衔接，关注室外花园、喷泉、花草和各房间之间的公共空间、大门入口平台、回廊等灰空间的处理和应用，提供一个优越的连接点，从视觉到心灵带给居者阔绰和通达的感受，让建筑和景观相映成趣，收放随意，动静皆宜，全面提升空间品质，体现生活场景和生活方式。

Reception Centre for Oriental Pearl, Taiwan China

Design Agency: Cayenne/Mulan Interior Design
Location: Taiwan China
Client: High-Yes Corp.
Area: 960m²
Photography: Cheng Xiangyi

中国台湾 "东方明珠"

设计公司：上海凯艳装饰设计有限公司
项目地点：中国台湾
客　　户：台湾海悦广告有限公司
项目面积：960m²
摄 影 师：程相怡

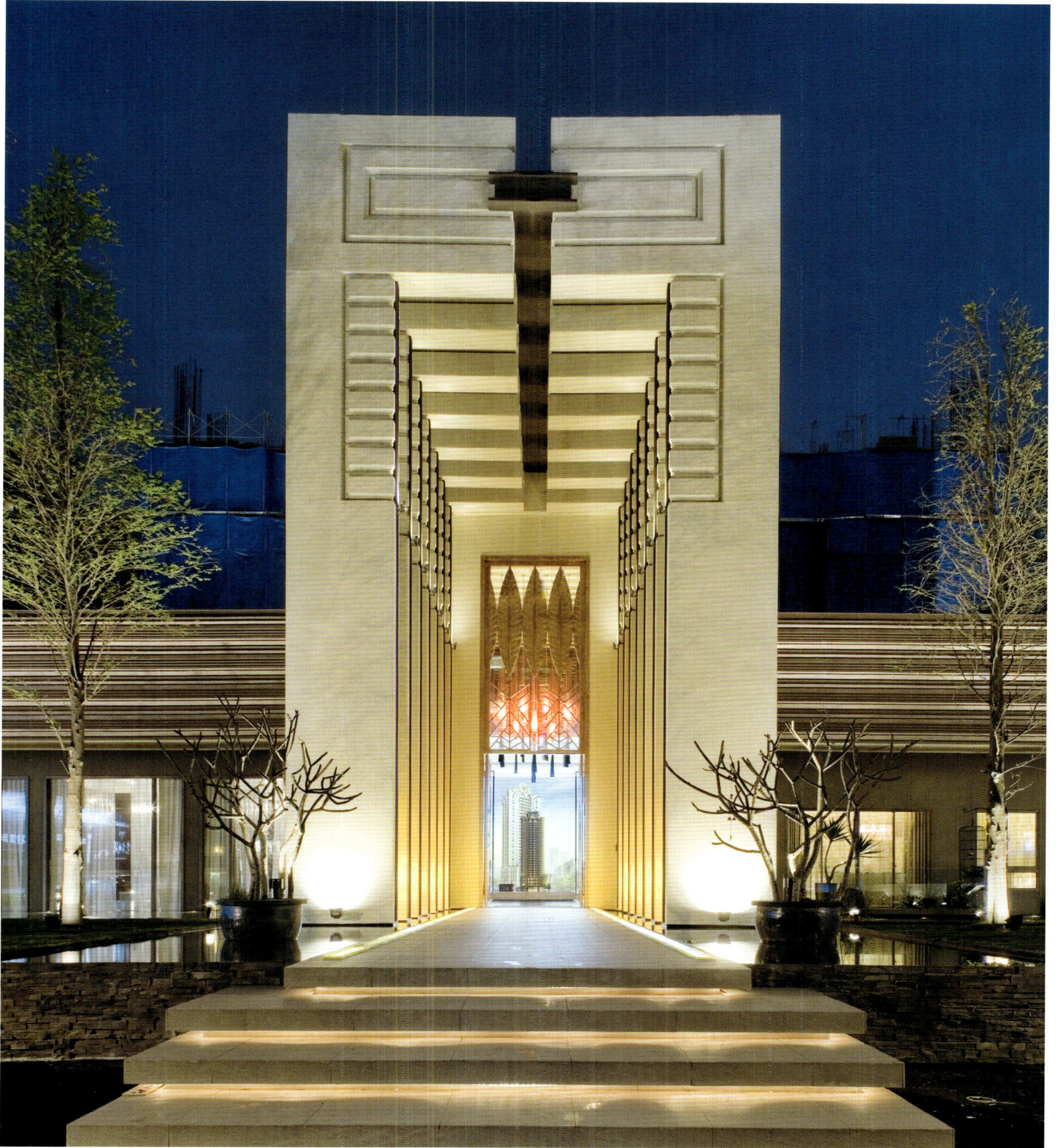

The project is the only steel-made transportation junction, a new model surpassing the design of Xin Yi Business Centre, a new model of 'East Manhattan', and a building as the City landmark. It reveals as itself an international Oriental Pearl.

The design of the entrance is steady yet creative, conveying the design concept of high fashion. The entrance is designed to be generous with strong visual impact, giving visitors a sense of everything new and fresh. Especially its unique shaping impresses every one. The entrance design well performs the function as guidance. The corridor is designed with the idea of Zen. The hundred-square-meter show house has an open pattern, which successfully interpret the bearings of a luxury residence. Its design blended in classical, contemporary, oriental and exotic styles, which shows its steadiness, elegance and superiority, and also its connection between contemporary urban society and royal court and its integration of architecture and natural environment. Everywhere one sees beauty, and everywhere presents beauty. The designer pays attention to life quality and details. Every element (lines, patterns) he uses is clear without any ambiguity. The designer's clear cut attitude attains the designing principle that "details are more important than everything".

This case hopes to pursue the leisure at the same time to reach the depth of culture, using the perceptual styles not only to illustrate the concept of leisure but also in an authentic way to express simple, nature, and fine leisure's culture. The style is simple without surplus modern design, presenting the gorgeous on the key parts. The design uses green environmental protection materials as much as possible and reduces the usage of the existing harmful substances such as paint, plywood. After the decoration, environment is very satisfying, which builds the idle environment and ensures the performance of environmental protection atmosphere just in the case of killing two birds with one stone.

全台湾唯一真正四铁共构的交通枢纽、超越信义商区的新板特区、新型东方曼哈顿，所有汇集于一身的地标性建筑，将展现成为国际间最耀眼的一颗东方明珠。

入口整体设计沉稳中带着新颖，高贵中透露着时尚的设计概念。入口设计大气、视觉冲击力强，使人耳目一新、眼前一亮，其造型独特，看过之后让人过目不忘，很好地起到了引导进入的作用。现代禅风的廊道设计，百平米样品屋有着开阔格局，成就豪宅气度。而其中的设计融合了古典与现代、东方与西方的对比形态。设计体现了庄重、典雅、尊贵，完成古典和现代的交融，现代都市和宫廷的联系，建筑和环境的自然融合，处处见景、处处是景。设计师注重生活品质和细节，设计中所采用的任何一个小件（线条、图案等）都清晰明确，毫不含糊，达到"细节重于任何事情"的设计原则。

本案希望在求索休闲的同时达到文化的深度，其以感性的手法诠释了休闲概念，地道地表现了质朴、自然、精致的休闲文化。简而没有多余的造型，却在关键部位渲出精彩，简洁而整体，大气而流畅。尽量采用绿色环保材料，控制使用油漆、胶合板等存在有害物质的材料数量，饰后空气环境非常令人满意，既营造了闲逸氛围又保证了环保性能，可谓一举两得。

Bosendorfer Reception Centre, Taiwan China

Design Agency: Cayenne/ Mulan Interior
Location: Taiwan China
Client: High-Yes Corp.
Area: 1,200m²
Photography: Cheng Xiangyi

贝森朵夫

设计公司：上海凯艳装饰设计有限公司
项目地点：中国台湾
客　　户：台湾海悦广告有限公司
项目面积：1200m²
摄影师：程相怡

The main element of this building's elevation is cloud. The world famous piano brand, Bosendorfer was created in 1828, Austria. It uses fabulous manual technology to create world's top pianos only for royal families, which made it the world's number one piano brand only for aristocrats. People who can own its pianos are the ones with noble social status. Hoping to become a "Bosendorfer" in the area of architecture, the luxurious residential space, Da Ping Shu international metropolitan mansion is located in the quite downtown, the centre of the Northern City, facing department stores, possessing convenient traffic. It seems that Da Ping Shu international metropolitan mansion is created as the symbol of energetic music.

On the top of its entrance, the circular decoration made Bosedorfer an architecture with vigor and power, successfully interpreting the idea of its name, Bosedorfer. Its exterior design uses gold copper color to create a luxurious spot light in the downtown area of the city. Its interior design uses the art of engraving. As a whole, this project is a perfect combination of fashion and art.

Bosedorfer is a project of hearing, sight, art and fashion. It uses mixed art to create a paradise. Its decorations and antiques increase the beauty of the space. At the same time it shows the spectacular feeling of the project, making it a symbol of identity.

建筑立面主要的装饰元素是云朵。

世界第一名琴"贝森朵夫"（Bosendorfer）品牌1828年诞生于奥地利，以精湛优异的手工技术打造仅供皇室贵族御用的顶级纲琴，传世至今占有无与伦比的地位，堪称为贵族而生的第一品牌，也寓意着只有最顶极的人士才能拥有它，而期许成为建筑业中第一品牌的"贝森朵夫"豪宅位于地段之冠、北市核心、"大坪数"国际都会豪邸，面对精品百货，交通方便，闹中取静。"大坪数"象征时尚的印象与动人的音符。

在入口大门顶的圆形装饰造就整个"贝森朵夫"无与伦比的气势，以奠定位于北市核心、媲美纽约第五大道的复兴北路上的"贝森朵夫"之亳宅霸气。在外观颜色上采用金铜色代表着令人叹为观止的奢华，成为精品地段的城市焦点，内部则以艺术雕刻的方式呈现，这就是时尚和艺术的结合。

贝森朵夫建筑本身就是一个听觉和视觉、艺术和时尚结合的精品之作，利用艺术的混搭，造就出一个人间仙境。而除了琉璃佛手的作品之外还有国际雕刻大师朱铭的太极系列和赵春翔的水墨画作等。这些陈设和配置着重于加强空间的欣赏性，同时也兼具了时尚的品位，呈现出精品的质感，更希望为住户打造层峰人士的品位及身份。

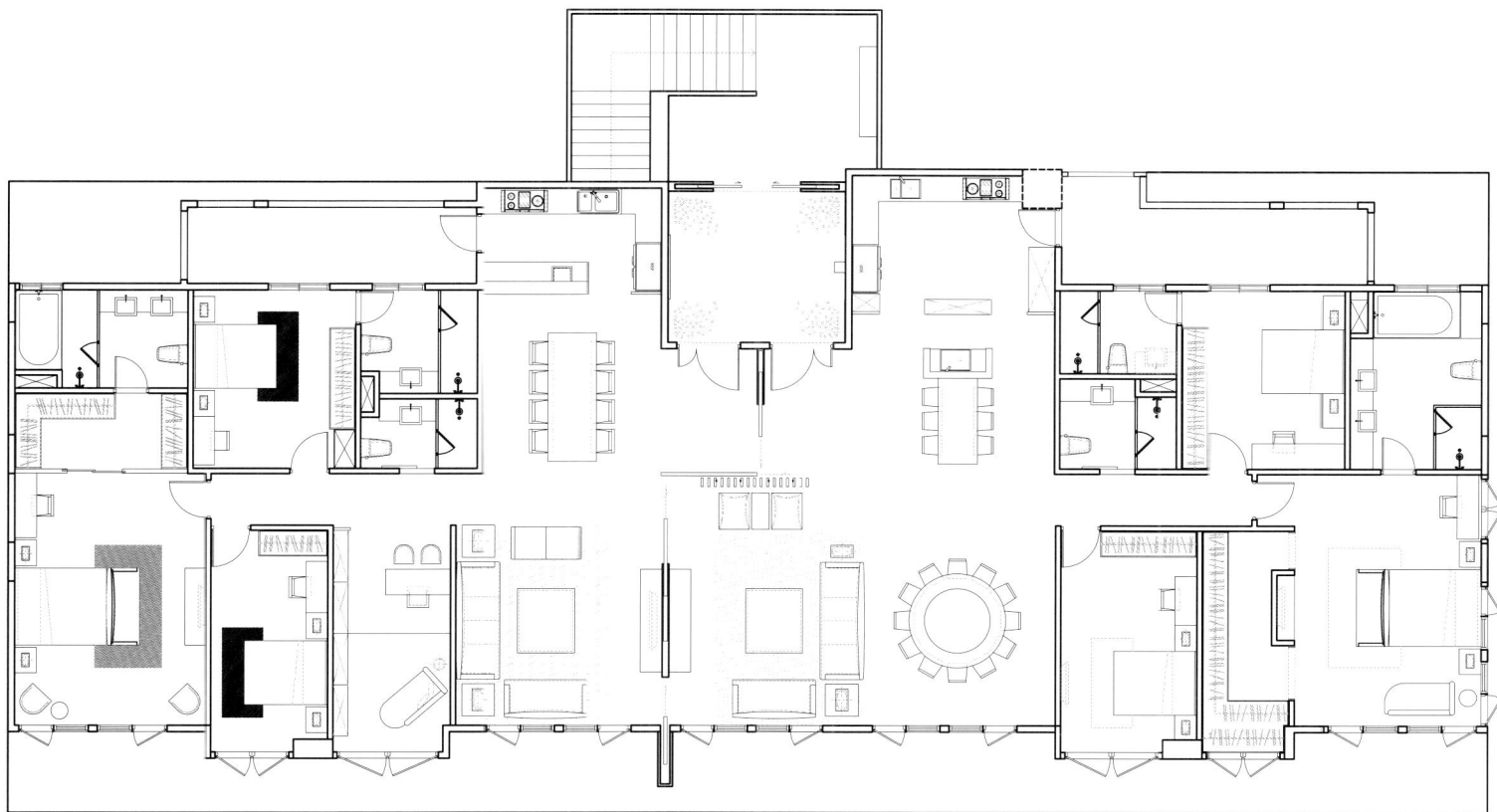

平面图一

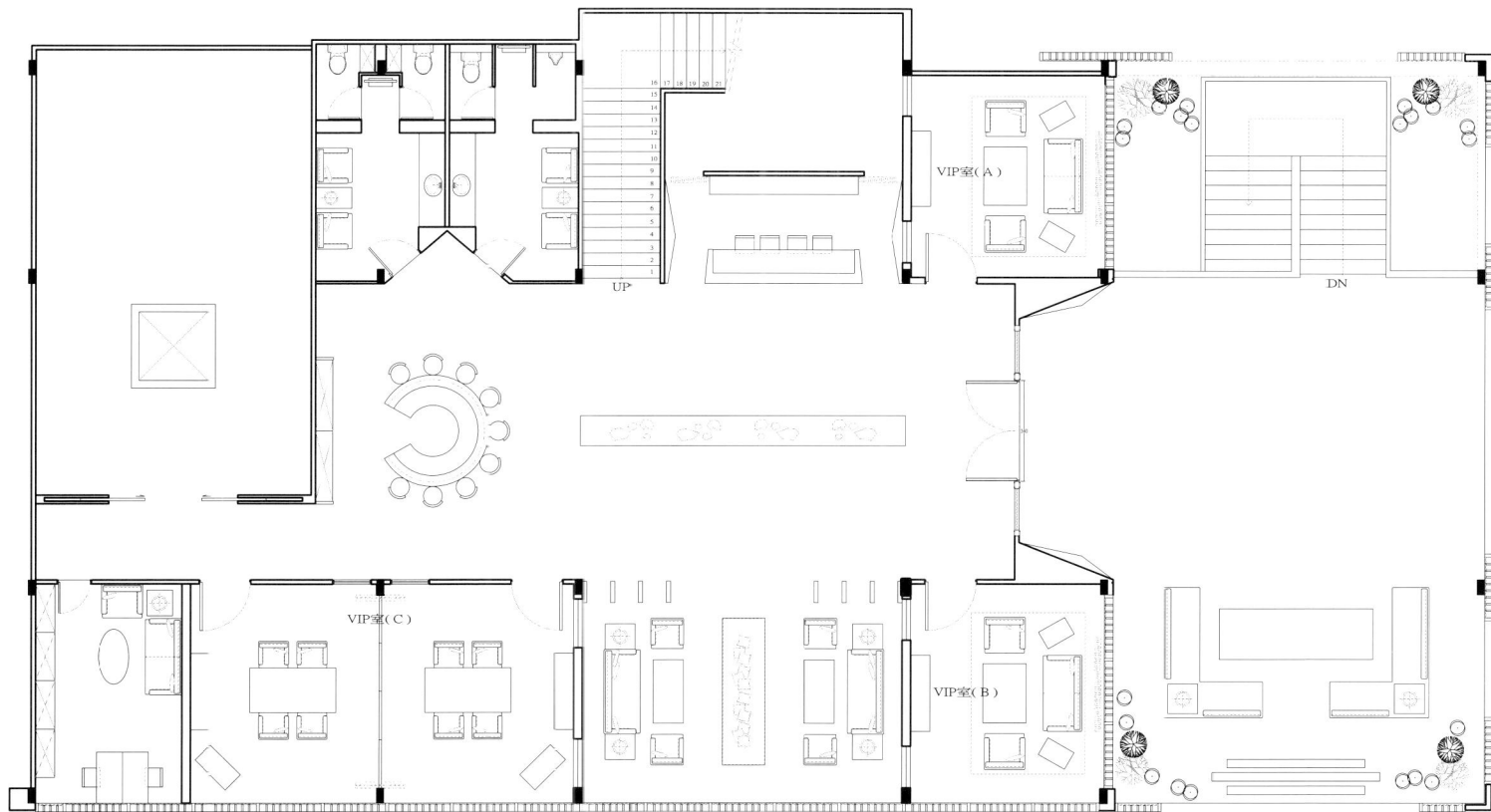

平面图二

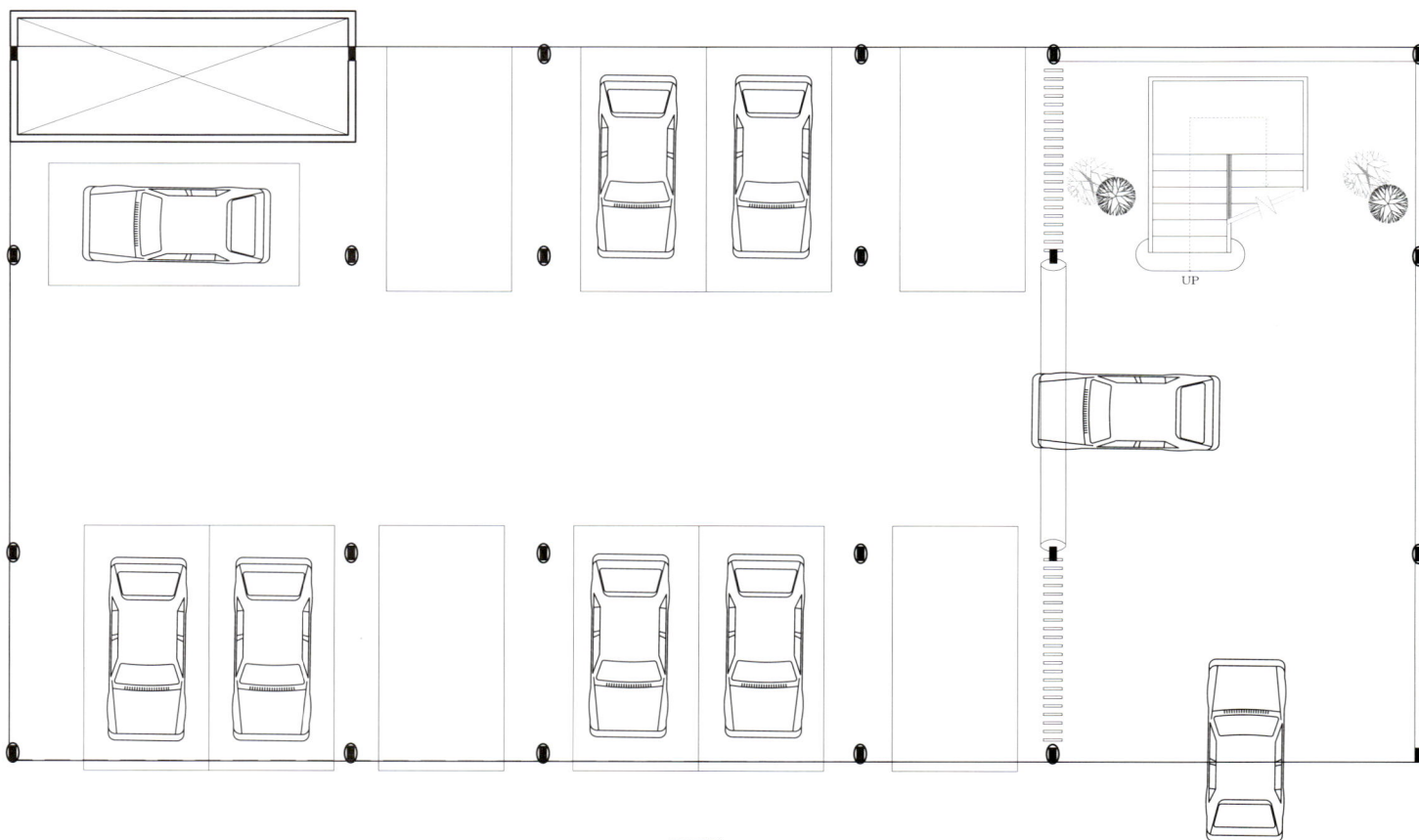

平面图三

Changhong Hongding Reception Centre

Design Agency: Cayenne / Mulan Interior Design
Location: Taiwan China
Client: High-Yes Corp.
Area: 1,300m²
Photography: Cheng Xiangyi

长虹虹顶接待中心

设计公司：上海凯艳装饰设计有限公司
项目地点：中国台湾
客　　户：台湾海悦广告有限公司
项目面积：1300m²
摄 影 师：程相怡

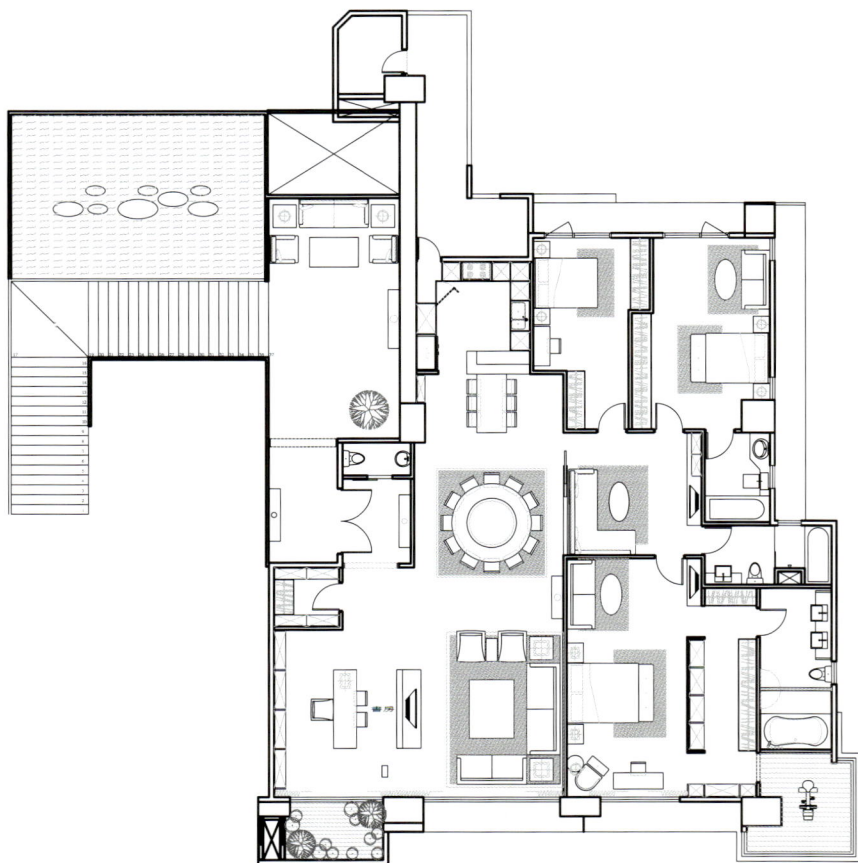

平面图一

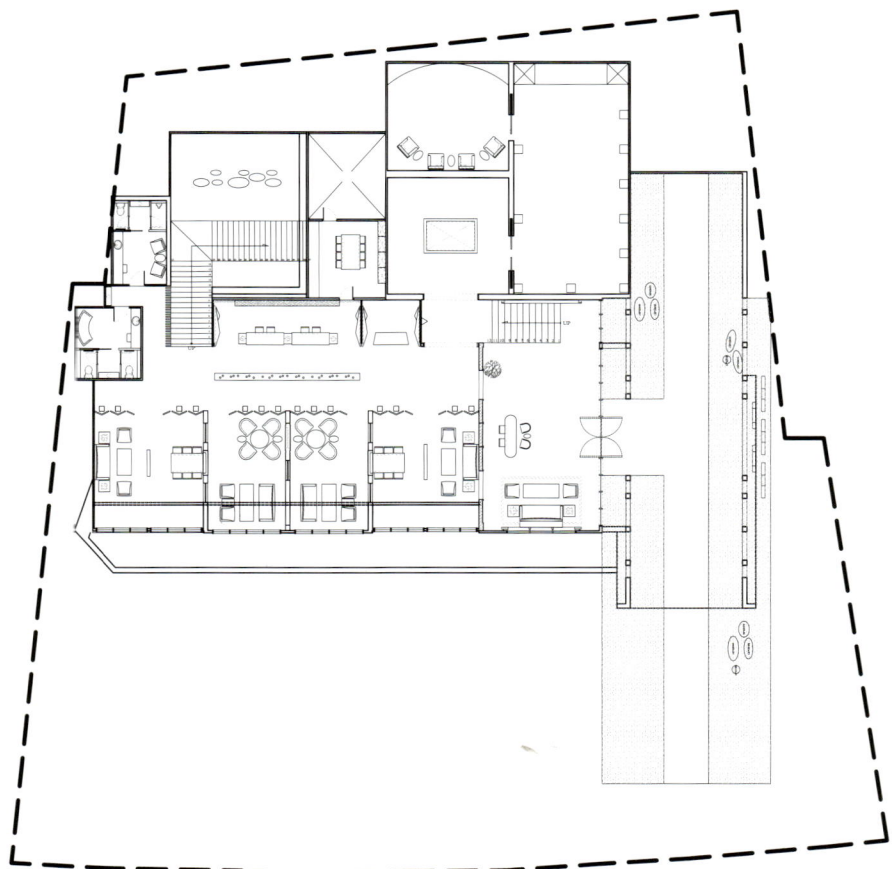

平面图二

The building elevation of this project is a mould of abstract geometric triangle, giving it a steady yet lively shape. Its interior design matches the architectural design, as to be restrained, steady, yet active. The point of the geometric triangle gathers visual effects, making its architecture and interior space a perfect match for each other.

When entering the reception centre from the entrance, with white-colored artistic decorations in the frameless pool on each side, one can sense the concept of icebergs. Every dawn, the white pebble-like artistic products glitter weak light, just like icebergs melting in the ice age, becoming small pieces of ice, spreading along the porch which connects the reception centre. It symbolizes the prelude of the mysterious Gold Triangle. The black wire and silver model is also one kind of artistic products. All of its designs and its technologies reveal one fact, that this architecture is a fabulous art. Together with the historical location, it sends out a message that this mansion has its own mystery, waiting for someone to discover.

Iron-craved flower lampposts are placed alongside the main road. Inside the lampposts, there are "falling leaves". This is a design combining special design and art, creating a relaxed atmosphere for the VIPs

The choice of lamps is very impressive. Every lamp is like an art product, combining beauty with usage. The design idea of the waterfall is to bring in nature. Also there are designs of wild geese symbols, "flying" among all the designs, and into the waterfall. There are also coordinating signs of blue wild geese on the wall of male restrooms, and red ones on the walls of female restrooms.

长虹虹顶的立面是非常抽象的三角几何造型，是一个沉稳但不失活泼的形状，室内空间跟随此几何形状呈现出内敛、沉稳主动的气氛，三角形尖顶部分更有凝聚视觉焦点的效果，故此建筑造型和室内空间是相辅相成、互相提升的。

从入口进入接待销售中心，两侧无边框水池里的白色艺术品呈现的是冰山的意境，在傍晚时，白色卵石般的艺术品会有白光从里面散发出来，有如冰山时期冰山溶解成一块一块的小冰山散布在两侧廊道引领着贵宾走进接待销售中心，象征着进入了神秘的黄金三角洲，揭开了神密宝藏的序幕。而黑色网丝和入口大厅的银色卧蚕式的造型也是造型艺术品的一种，代表此建筑是设计、精工都简单有力的沉稳之作，再配合上此建筑以人文历史为背景的官邸地理位置，处处都显示出这个自然流动的艺术人文大宅有着等待您发掘的神秘感。

主要通道两侧所放置的是制铁仵刻花灯柱，灯柱上刻花的元素是流动中的落叶，是空间设计装置同时兼为艺术品，从落叶镂空中透出晕光，点缀出如同在树阴底下轻松悠闲的氛围，样式不浮夸又没有多余的装饰，简单重点式地带领贵宾通往贵宾专属的VIP室。

整个空间内选用的灯具令人印象深刻，每一件都像艺术品，把唯美的造型与照明的功能完整地结合在一起。其中有一组灯具像一群大雁在林间飞舞，飞入了水瀑，进而散布于各区域，将自然的氛围带入室内。在卫生间的入口也以有颜色的落雁分男卫生间（蓝色）和女卫生间（红色）与之相呼应。

XYI Design

隐巷设计顾问有限公司

黄士华

Just as the 'Yin Xiang' literally, a low-key, practical, simple concept, an innovation of 'every cloud has a silver lining', less, but better. X and Y respectively represent the first letter of 'Yin' and 'Xiang', which also act as the two corresponding axes of space and extend the limitless possibilities. I stands for 'INTERIOR'.

Founded in Taipei in 2010, and the same year in July in Qingdao, China, Zang Nong Interior Design Ltd., as a branch unit in Mainland China, was established and responsible for operation in Mainland China, which mainly engaged in various space planning and design, furniture design, design consultants, etc. It shortlisted the 2010 TID single household space design awards.

如"隐巷"字面般，一种低调、实务、质朴的理念，一种柳暗花明又一村的创新，少而精。XY各代表"隐""巷"拼音字首，XY也为空间中两个相对应的轴线，可延伸出无限的可能性，I为INTERIOR。

2010年成立于台北，同年7月于青岛成立藏弄室内设计有限公司，为大陆区分支单位，负责大陆地区营运，主要从事各类空间规划设计、家饰设计以及设计顾问等工作。

入围2010 TID单层居家空间设计奖。

The Orient Golden Stone Reception Centre

Design Agency: XYI Design
Location: Qingdao
Area: 1,400m²
Photography: Lu Zhenyu
Main Materials: gold foil paper, silver foil paper, 4mm solid stainless steel circle, mirror finished stainless steel, electroplated panel, 6mm tawny hollow plate, 20mm paint density board sculpture, walnut, gold off white marble, 4mm copper pipe, rusting copper plate, imported crystal glass, laminated glass, metal mosaic, cafhide leather

东方金石接待中心

设计公司：隐巷设计顾问有限公司
项目地点：青岛
项目面积：1400m²
摄影师：卢震宇
主要材料：金箔纸、银箔纸、4mm实心拉丝不锈钢圈、镜面不锈钢、电镀板、6mm茶色中空板、20mm 烤漆密度板雕刻、胡桃木、黄金米白理石、4mm 铜管、锈蚀铜板、进口水晶玻璃、夹胶玻璃、金属马赛克、小牛皮皮革

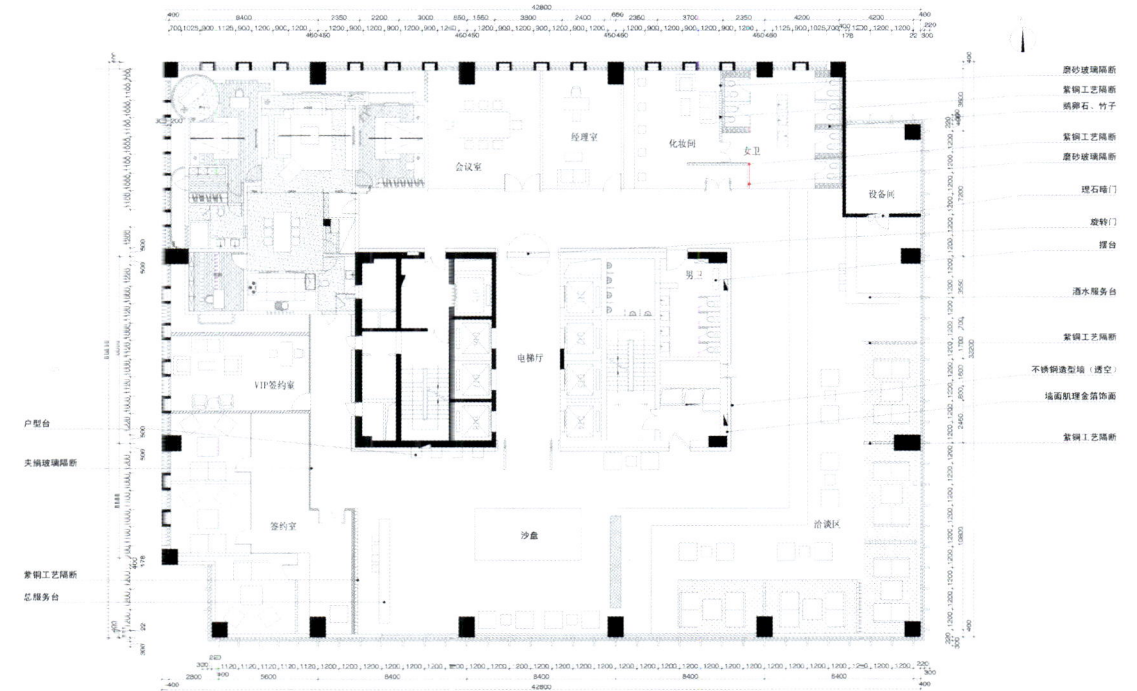

平面图

The client of the case has mining industry background, and intends to make the best reception centre.

There was a large scale Core Cylinder in the building. We have tried various materials on it but none of them could create a sense of luxury ore. So we decided to cover it with some special painting, and then extrusion texture, finally put gold foil paper on it to indicate the concept of Golden Stone directly.

The main materials used in this case was some modern materials, such as, continuity circle element created the sense of the Orient, the contrast of rough and delicate, bright and dark, traditional and modern, all of these created a strong space feeling. The use of various materials enhanced the fusion of Golden Stone and the Orient concept.

The main generatrix was separated by Crystal Chandelier. The flare that produced when the light went through the Crystal increased luxurious atmosphere. The original wall of the elevator lobby was made into a flexible feature wall. The existence of the left wall was to create different gradation of the space and it also acted as the information wall, while the right side was reception area.

Many compartments that made of copper pipes and made into plant shape were used in this case, both beautiful and taken care of the privacy. The floor of the negotiation area was raised. Hence, the client could enjoy the amazing seascape while sitting on the couch. The space function was separated by different floorings. And also there was a bar area. The reception centre could become a club in the future.

本案客户具有矿业的背景，希望能打造出胶南海湾最好的接待中心。

建筑体有一大型核心筒，想藉由核心筒形成具有体量感之造型，经尝试，许多材料均无法呈现奢华矿石之感，故采用金箔面料，先于墙面上覆上特殊涂料，压出纹理，最后再覆上金箔纸，直接明确地表现出"金石"概念。

整体设计材料主要以现代感之材料构成，藉由连续性圆圈元素，塑造东方感，利用粗矿与细致、明与暗、传统与现代之对比，产生巨观之空间感，藉由混合材料的形式强化金石与东方概念融合。

主要动线以水晶玻璃吊灯作为区隔，灯光透过水晶玻璃产生之炫光增加奢华感，原有两侧信道之电梯厅，制造一座活动式端景墙。左侧之墙体为使空间产生层次感而存在，同时也是信息墙，右侧为服务接待区。

此案采用许多非遮蔽性之隔间，由铜管制成之隔间，放入植物的造型、美观与隐私兼顾，并融合中式竹林造型。洽谈会客区将地面抬高，为了让客户坐在沙发上同样能欣赏海景，透过不同的地面材料界定空间功能，并配置一吧台区，能同时兼有宴会、餐饮之功能，未来接待中心可转型为会所。

Vipassana Design Consult Organization

内观设计咨询机构

赵力行

Zhao Lixing, born in 1972, Design Director of Neiguan Design Agency, and the standing director of the third session of council of Shenzhen Interior Designer Association and won many prizes in national and international design competitions and credits in 'Shenzhen Design Thirty Years Thirty people'. Since graduated from Hubei Institute of Fine Arts, 1996, majored in design, with his specialized qualities and the enthusiasm over design career, he made his own contribution to the design career of China. And he will carry on...

1972年出生，内观设计咨询机构设计总监，深圳市室内设计师协会第三届理事会常务理事，多次获得国内外设计竞赛奖项，获评"深圳设计三十年三十人"的荣誉称号。自1996年由湖北美术学院设计系毕业至今，凭借良好的艺术专业素养及对设计事业的持续热情，不断探索，希冀能够探索出一条将深厚的东方传统文化底蕴与当代生活模式自然结合的设计语言体系，为中国的设计事业贡献一份绵薄之力。在路上，用心走……

Huizhou Golden Bay Sales Centre

Design Agency: Vipassana Design Consult
Organization
Location: Huizhou
Client: Financial Street in Huizhou
Area: 420m²

惠州金海湾销售展示中心

设计公司：内观设计咨询机构
项目地点：惠州
客　　户：金融街惠州置业
项目面积：420m²

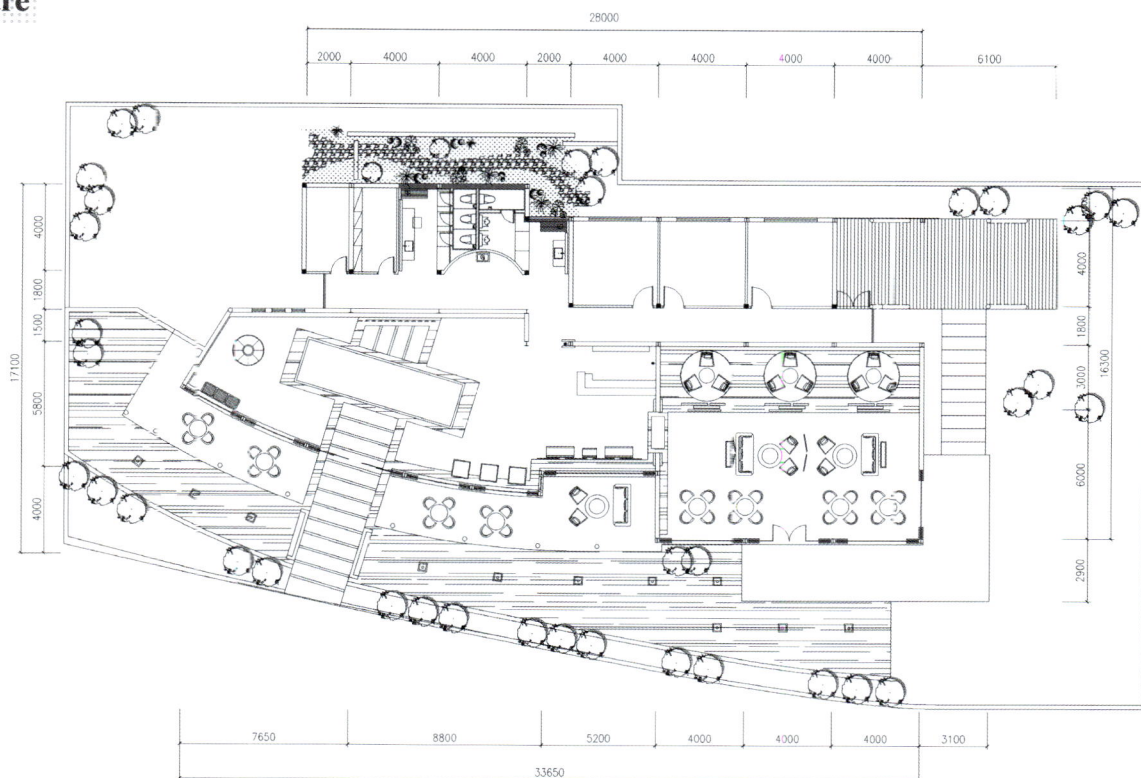

平面图

As the first-stage project sales centre of the largest waterfront estate in southern china of Beijing Financial Street Real Estate, the overall floorage of this sales centre is 420 square meters, consisting of project exhibition area, negotiation area and the logistics office. Designer took 'enjoy the leisure coastal lifestyle' as a point of penetration, using modern design language and local architectural and humanistic elements, creating a space that providing people with comfortable club-like service. Doing business while enjoying coffee, fruit and sunshine...

作为北京金融街地产在中国南部最大的滨海地产项目一期销售展示中心，此案建筑面积420m²，分为展示区、洽谈区及后勤办公区三个区域。设计师尝试以体验优质的滨海休闲生活方式作为切入点展开设计，以"海湾——世界又一滨海胜地"广告推广语的气度为诉求，采用现代手法，吸取当地建筑及人文元素，消解"销售"元素，强化"享受"体验，以会所式的亲和服务，让整个项目展示及销售在一杯咖啡、一碟生果、一小段明媚阳光之间惬意地完成……

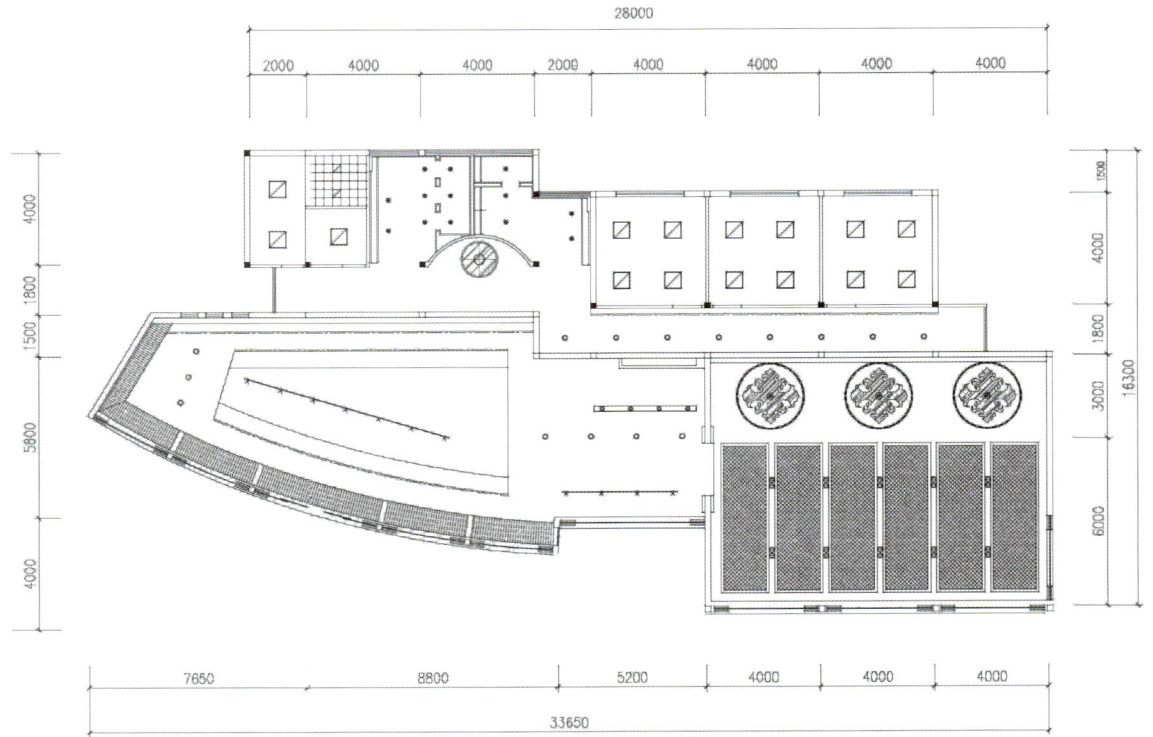

天花图

Golden Triangle Sales Centre

Design Agency: Vipassana Design Consult Organization
Location: Tianjin
Client: Tianjin Century Estate
Area: 260m²

金三角销售展示中心

设计公司：内观设计咨询机构
项目地点：天津
客　　户：天津世纪地产
项目面积：260m²

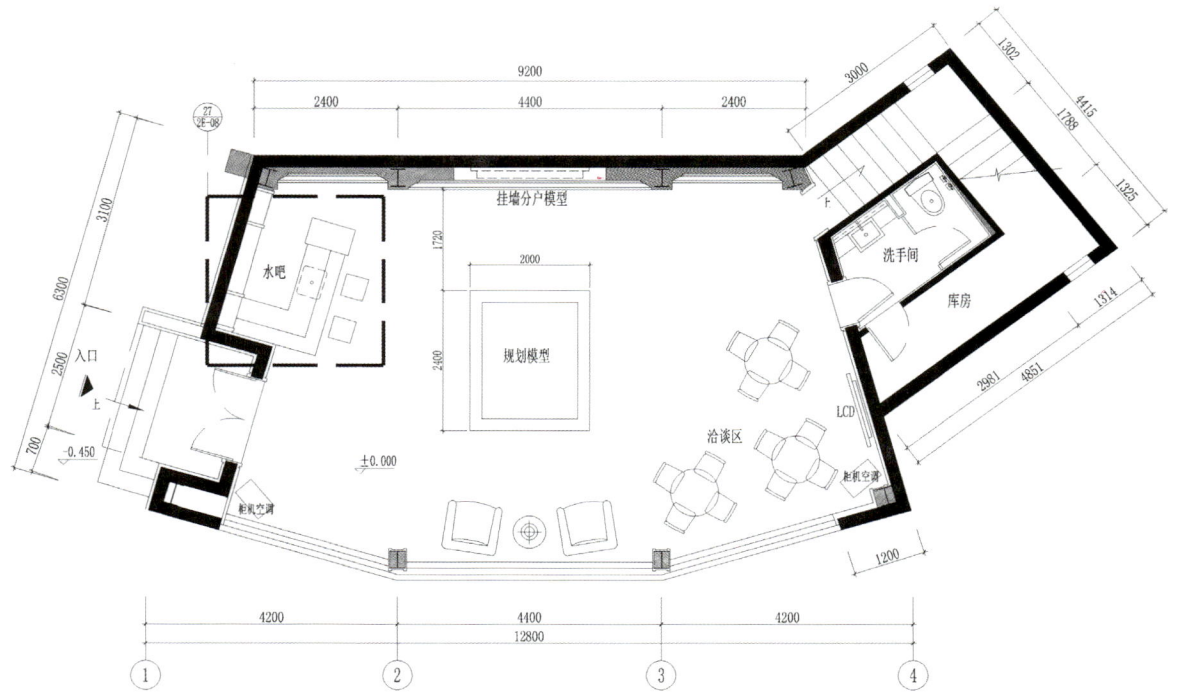

1F平面图

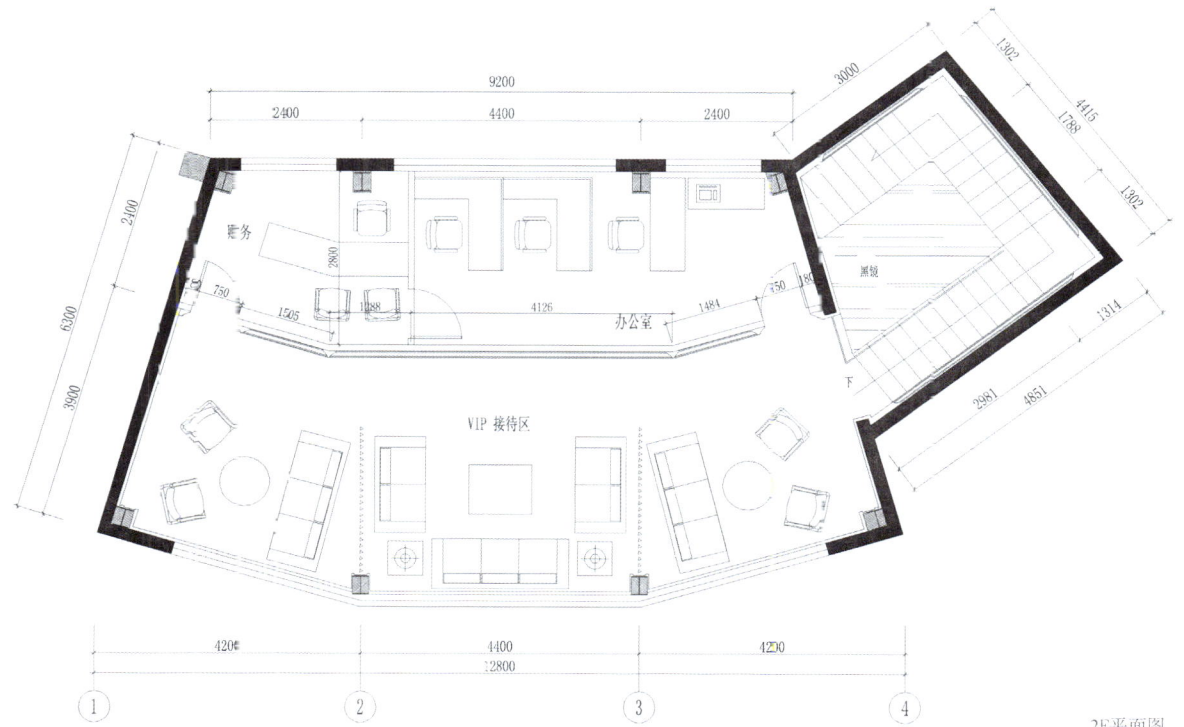

VIP 接待区

办公室

商务

照镜

2F平面图

As a high end residential project in Binhai New Zone, Tianjin City, to reasonably allocate the space as well as to reflect project quality and to satisfy the client who wants a short construction period, is a great challenge. At the same time, the style of the sales centre must go with the whole project positioning, so the simple European style is a proper choice. Consider of the construction period limitation and tense space, we reduced the work of carpentry and painter and used unified design of the ceiling. There were large area black mirror and light box set on the wall to promote construction and improve the atmosphere and quality of the sales centre. Hence, after twenty days of construction, a small but delicate sales centre appeared in front of everyone.

作为天津滨海新区钻石地段的高端住宅项目，在寸土寸金的有限使用面积内，如何合理分配空间、同时体现项目品质，是一个有趣的挑战，同时还需要应付业主需求的极短的施工周期，风格必须尊重项目的整体定位，简欧是合适的选择，考虑到工期限制及紧张的空间，减少木作及油漆工程量，天花采用统一的设计，墙身设置大面积黑镜及灯箱，扩充空间，强化售楼处氛围及品质，这样在二十天施工后，一个小而精致的营销展示空间就呈现在大家面前。

Zhongjian Huafu Sales Centre

Design Agency: Vipassana Design Consult
Organization
Location: Dezhou, Shandong
Client: Zhongjian Real Estate
Area: 1,800m²

中建华府销售展示中心

设计公司：内观设计咨询机构
项目地点：山东德州
客　　户：中建地产
项目面积：1800m²

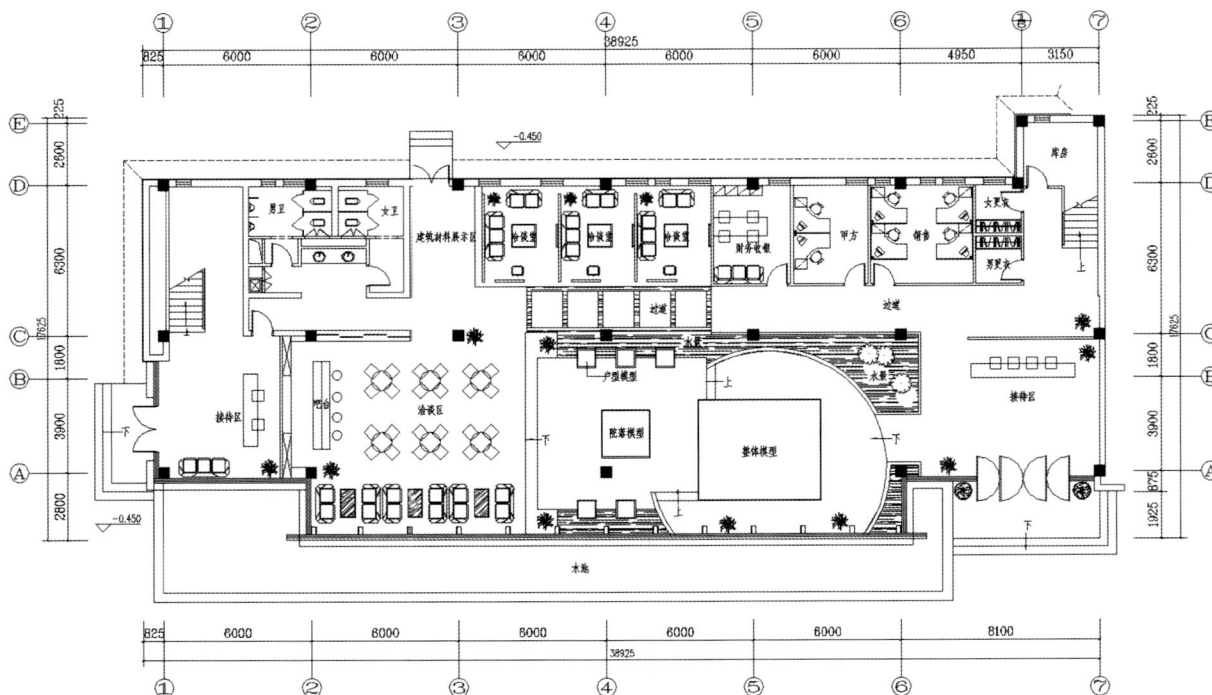

1F平面图

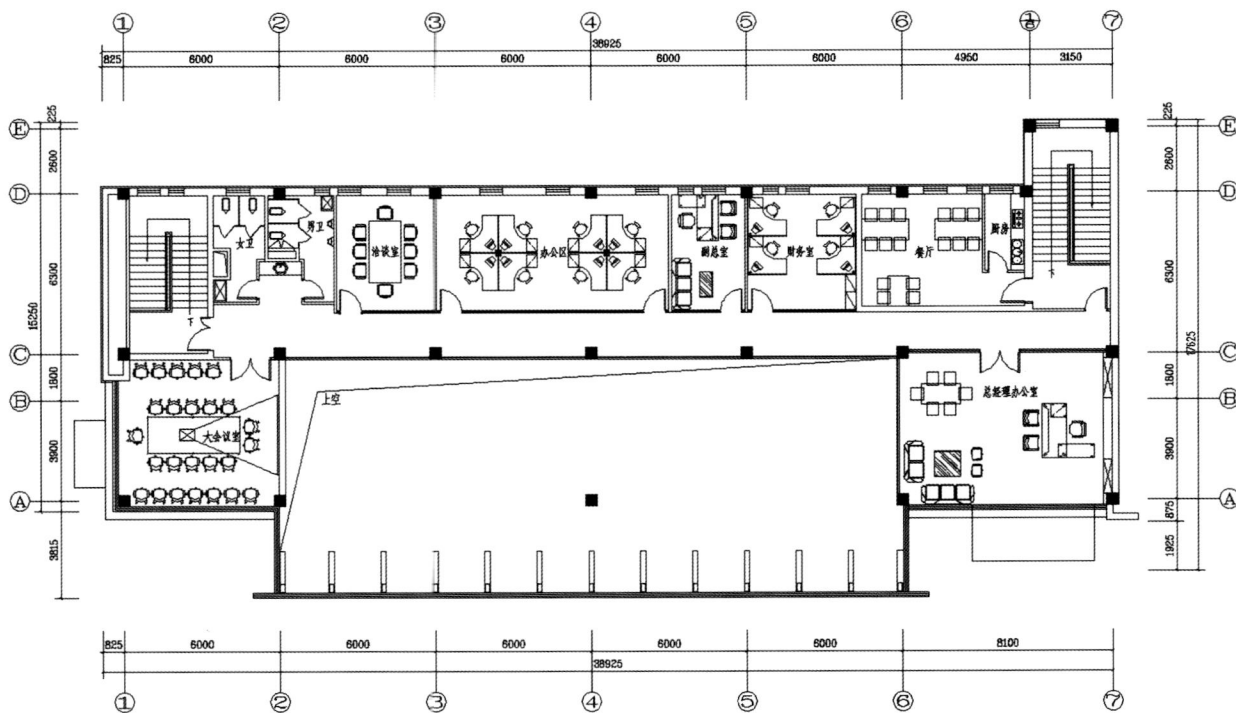

2F平面图

This project is located in Dezhou, Shandong Province. Its overall floorage is 1,800 square meters, consisting of entrance area, project exhibition area, negotiation area, signing area and the logistics office. According to its fine building structure, using simple design language the designer created a great sales centre.

本项目坐落于山东德州，建筑面积1800m²，分为入口接待区、项目展示区、洽谈区、签约区、后勤办公区。依据建筑良好的空间条件，室内设计以清晰的动线组织、合理的空间关系、简洁的设计语言，营造一个具有东方优质生活品质的空间、内敛而气韵悠长的销售展示场所。

Deep Design Consultanting

厦门宽品设计顾问有限公司

李泷

Fuzhou University , Arts and Crafts College,
Department of Environmental Art Design
Member of The China Institute of Interior Design
Council Member of Asia Pacific Federation of
Architects / Interior Designers
Design Director of Xiamen Deep Design Consultanting
Limited Company

Beijing Shanshuiwenyuan Club, Jiake Group(China)
Headquarter, Chuanguan Group (Hong Kong)
Headquarter, Gulangyu Nazhai Boutique Hotel, CNPC
Fujian Headquarter, Changsha Yosemite, Beijing
Wanchenggongguan Sample House, Guanzhi Spring
Hotel, Guanyin Mountain International Business
Centre

福州大学工艺美术学院环境艺术设计系
中国建筑学会室内设计分会会员
IAI亚太建筑师与室内设计师联盟理事会员
宽品设计顾问有限公司设计总监

主持专案
北京山水文园会所
佳科集团（中国）总部
创冠集团（香港）总部
鼓浪屿那宅精品酒店
中石化福建总部大楼
长沙优山美地
北京万城公馆样板房
冠豸山温泉度假酒店
观音山国际商务营运中心

Changsha Yosemite Reception Centre

Design Agency: Deep Design Consultanting
Location: Changsha
Area: 800m²
Main Materials: white kageki, ariston white marble, mosaic, paint coated glass

长沙优山美地接待中心

设计公司：厦门宽品设计顾问有限公司
项目地点：长沙
项目面积：800m²
主要材料：白影木、雅士白大理石、灰木纹马赛克、烤漆玻璃

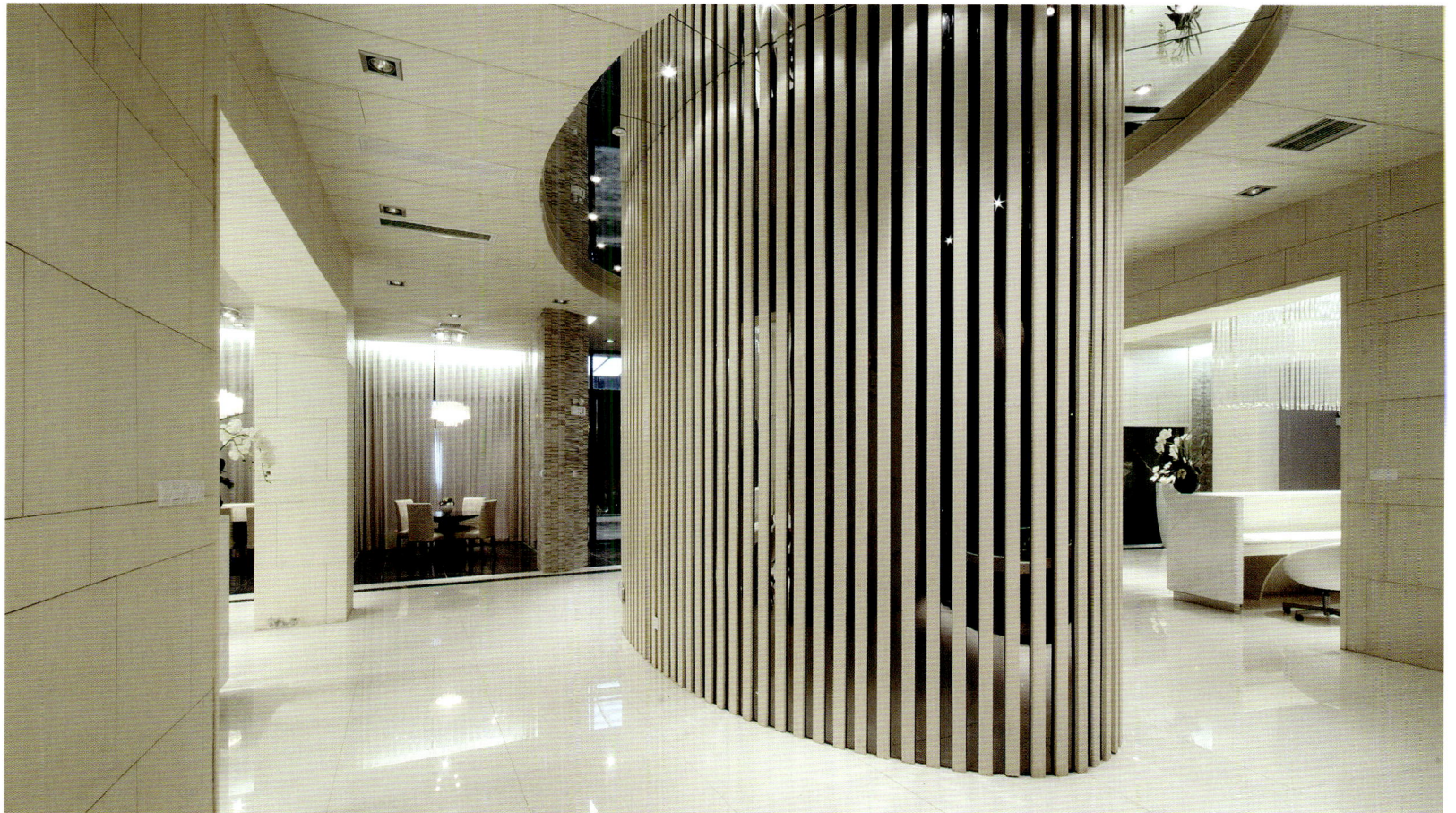

This project aims to show the advocated lifestyle by arousing the people's expectation with the beauty of this project, and then finally promote the sales.

The restriction of architecture structure is the biggest challenge at the beginning. The priority of the project is to figure out the way how to build an orderly atmosphere within the messy structure. The structure becomes symmetrical after elaborate planning. The enormous construction columns in the original space centre are surrounded into a special space which becomes the eye focus. The five-meter height oval shape structure is magnificent; what's more, it connects each function space with its fluent shape.

Milky white is the main color tone of the whole architecture, while the integrating materials manifest the elegant atmosphere. The funny spatial structure and elegant atmosphere weaken the sense of the commerce feeling of the project, and boost the easy communication between guests. All the above show the quality of the project.

The designers try to find the origin of the project by simple techniques. In the end, the designers make the structure more idealistic by using simple and pure materials and sprightly professional skills with the purpose to find the balance between commerce and art, at the same time to find the resonance between house owner and the project.

本案的重点意在展示项目所倡导的生活方式，以唯美的意境激发受众的期待，从而促进项目的推广与销售。

项目建筑结构的局限性是设计初期最大的挑战，如何在凌乱的结构体中构筑空间的秩序和气势是方案的设计重点。经过精心规划后的空间呈对称格局，原建筑中心的庞大结构柱被围合成空间的视觉焦点，以圆润的椭圆造型呈现，近五米高的卵形体量营造非凡气势，并以其特有的流畅性组织各个功能空间的相应关系。

柔和明亮的米白色系是空间贯穿始终的主色调，统一和谐的材质搭配营造项目精致典雅的高品位形象。流畅而具有趣味感的空间规划，亲切而优雅的环境塑造，淡化了项目的商业氛围，促进愉快而轻松的交流，充分展示出项目本身内在的气质与品位。

设计师在定位中试图通过简洁、温馨的设计语言来寻找和陈述设计的本源，捕捉空间的情绪。本案运用通透纯净的材质及简约明快的造型手法赋予空间理想主义的色彩，在商业与艺术之间找到了平衡点，在业主与项目之间找到了共鸣。

Fuzhou Zuoze Decoration Engineering Co.,Ltd

福州佐泽装饰工程有限公司

王添春

Employed qualifications: Zuoze Decoration Engineering Co., Ltd
Deputy general manager / deputy chief designer
Association of Chinese interior decoration members
China building decoration interior architects
2007 Fujian architectural decoration industry outstanding young designers
2007 Fuzhou top new interior designer
Fuzhou craft art technology school art design speciality visiting senior lecturer

Engaged in interior design for 10 years, specialized in public spaces and luxury villa design, win Fujian interior design contest for two consecutive years.

Accept project interviews of "Strait metropolis daily", "Fujian home decoration online website", "Fujian three sets of and Fuzhou three sets of (home in Fuzhou) columns" and "Fujian economic channel (home buyers square) columns" many times.

Many works are included in "2006 Fujian Tiannuo cup of indoor and environmental design grand competition works" , "Household new claims" etc.

从业资历: 佐泽装饰工程有限公司
副总经理 / 副总设计师
中国室内装饰协会会员
中国建筑装饰室内建筑师
2007年度福建省建筑装饰行业优秀青年设计师
2007年福州新锐室内设计师
福州工艺美术技术学校艺术设计专业客座高级讲师

从事室内设计十年，专攻公共空间及别墅豪宅设计，连续两年获得福建省室内设计大赛一等奖；
多次接受"海峡都市报""福建家装在线网站""福建三套和福州三套(家在福州)栏目""福建经济频道(置业广场)栏目"专题采访；
多项作品被收录在《2006年天诺杯福建省室内与环境设计大奖赛作品集萃》《家居新主张》等图书和杂志。

The bund 1st Sales Office

Design Agency: Fuzhou Zuoze Decoration
Engineering Co., Ltd
Location: Fuzhou
Area: 500m²

外滩1号售楼部

设计公司：福州佐泽装饰工程有限公司
项目地点：福州
项目面积：500m²

The life of sales office is short, but its mission is extremely important to developers. The question of designers should emphatically consider is how to spend the least money to achieve the best effect. We just want to use low cost to make an interesting space, and build a high-grade atmosphere which is close to the trend of the time pace in this case.

Quiet and elegant fresh peak green, beige and white are used as the main colour element of the space. Besides, plant imagery is reflected by modelling elements, such as the droplight of branches modelling, color of negotiate district seat, potted plants, all reflecting fresh breath of nature.

The design style strives to be conciseness, lively, natural and generous and when you walk into it, you'll feel bright. There is a kind of strong vitality and dynamism in your heart. It's not luxurious but very modern and engaging. What's more, it doesn't pursue high grade in material, but practicality, reflecting the elegant style.

　　售楼部的生命是短暂的，而它的使命对开发商来说却是极其重大，如何用最少的钱达到最佳的效果，是设计师着重要去考虑的问题。在本案里，我们就是要用低廉的成本打造一个有趣的空间，营造一个紧贴时代潮流步伐的高品位氛围。

　　在色彩上，以淡雅清新的嫩绿色、米黄色、白色作为空间最主要的色彩元素。另外，造型元素多体现植物的意象，比如树枝造型的吊灯、洽谈区座椅的颜色、植物盆栽，都体现出了大自然清新的气息。

　　本案的设计风格力争简洁、明快、自然、大气，并且要有给人走进去心头一亮、给人活力和冲劲的感觉，不豪华但非常现代、耐看，并且在用材上不追求高档，但却实用、实惠，并体现其高雅的格调。

洪德成设计顾问有限公司

洪德成

Master of Business Administration in Milan, Italy,
Institute of Technology Design College
Member of HK IDA Interior Design Association

Hong Decheng has engaged in interior design since 1980 and has 20 years' professional interior design experience. He is a pioneer of Chinese interior design as well as a witness of thirty years of interior design. In 1988, he began his personal career, and has applied himself to research and development of interior design since then. Afterwards, in 2001, he established Hong Decheng Design Consultancy HK Co., Ltd, and then in 2008, he established DHA International Design Consultant Institute with Italian designers specializing in providing high qualitative professional services of interior design for trade investment companies as well as customers worldwide.

Hong Decheng has been concentrating on annotating spatial art in a unique and professional perspective leading to developing a new style and creative pleasure. He works hard in forming inspiration and elements in daily life and putting them into works. He regularly holds the construction of furniture, works of art, lighting atmosphere. Therefore, his details oriented works can always touch people with unique feature.

意大利米兰理工学院设计管理硕士
香港IDA室内设计协会会员

　　洪德成自20世纪80年代以来一直从事室内设计工作，拥有超过20年的专业室内设计经验，是中国室内设计的先驱者与开拓者，也是中国室内设计30年发展的见证者与亲历者。他于1998年开始发展个人事业，随后一直致力于建筑室内设计的研究与发展，2001年创办洪德成设计顾问（香港）有限公司，2008年与意大利设计师共同创立DHA国际设计顾问机构，专为世界各地的商业投资公司及客户提供高品质的建筑室内设计专业服务。

　　洪德成一直专注于以独特的专业视角去诠释空间艺术，每次都能独辟蹊径，令人耳目一新。他在生活中不断提取设计灵感和元素并巧妙运用于作品之中，对家具、艺术品及灯光氛围的营造亦张弛有序，他对空间细节的琢磨近于苛刻，因而所表现的独特之处常能打动人心。

Dayun City-States Exhibition Hall

Design Agency: Hong Decheng Design
Consultancy HK Co., Ltd
Location: Shenzhen Yitian Holiday Square
Area: 154m²
Main Materials: black mirror, clear glass, self-
leveling, red&white piano paint, ICI

大运城邦展厅

设计公司：洪德成设计顾问（香港）有限公司
项目地点：深圳市益田假日广场
项目面积：154m²
主要材料：黑镜、清玻、自流平、红色及白色钢琴烤
漆、ICI

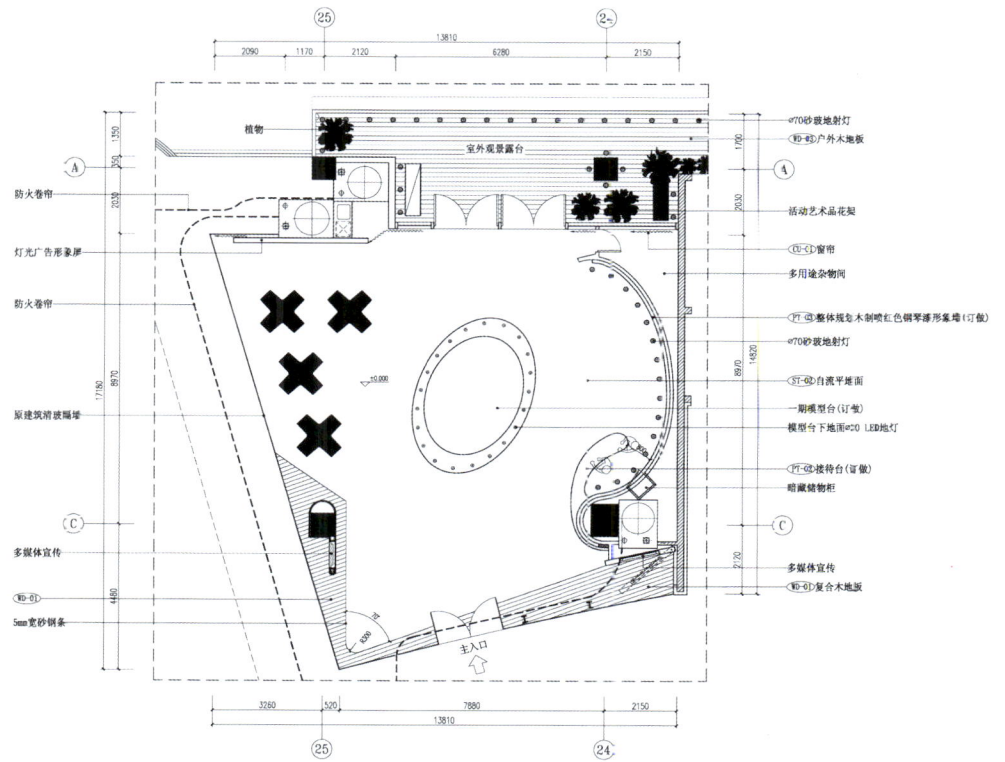

展厅平面图

Dayun City-States Exhibition Hall is located in the Shenzhen Yitian Holiday Square. It's mainly used for advertising, displaying and selling consultancy of the Dayun project of Yitian Holiday Square. Consequently, it's different from other house sales offices. At the same time, as a high-end commercial centre, Yitian Holiday Square is originally a good place for displaying and selling various high range products. It has become a difficulty of design to stand out from all kinds of competitors and to highlight the design style of the hall which also raises more challenges for designers.

Finally, designers used bold design elements of luxury yacht as the theme of the exhibition hall design. It runs through the whole hall design with changeable arcs, and breaks through the limitation of space, producing the fluent innervations of the whole space. It has shown up the theme of sporting of the Dayun project. The white cabin shape building model is like a sailing yacht, floating on the vast sea. In the gey and white, it gives priority to the exhibition space. An arc setting wall with eye-catching red adornment anaglyph became another centre vision. Designers extremely fuse the regional model into decoration. The massive modelling strewn at random in the illuminating of spotlight sends out an attractive lighting effects, giving rise to a strong visual impact. The Dayun City-States Exhibition Hall attracts sight of passers-by as soon as its completion and it becomes a memorable scenery within Yitian Holiday Square.

大运城邦展厅位于深圳市益田假日广场内，主要用于益田集团大运城邦项目的宣传、展示及销售咨询，因此不同于一般的楼盘售楼处的设计。而益田假日广场作为高端商业中心本身就是各类高档商品展示及销售的场所。如何在众多的卖场中脱颖而出、如何突出展厅的设计风格，这也成为大运城邦展厅的设计难点，也为设计师提出了挑战。

最终设计师大胆采用来自豪华游艇的设计元素作为展厅设计的主题，以多变的弧线贯穿整个展厅设计，突破空间约束，使整体空间产生流畅的动感，突出了大运城邦项目"运动"的主题。白色船舱造型的楼盘模型台，像一艘即将远航的游艇，漂浮在广阔的海面之上，成为整个展厅设计的焦点。在以灰色与白色为主调的展厅空间之中，弧形背景墙面上夺目的红色装饰浮雕成为又一个视觉中心，设计师极具巧思地将楼盘区域模型融合在浮雕装饰之中，错落的块状造型在射灯的照射之下，散发出迷人的光影效果，造成强大的视觉冲击力。大运城邦展厅一经落成，即以独特的视觉设计吸引住过往行人的目光，成为益田假日广场内又一道令人难忘的亮丽景观。

深圳大羽营造空间设计机构

冯 羽

A designer should lead the market instead of being assimilated by the market. Great perseverance, strong belief and martyrdom will are all necessary.

Interior design is by no means decoration design. It is a spatial experience based on the designer's rich cultural background, or rather, on his sentiments towards space.

I like things that come naturally. A space which rose out of sincere emotion, without high technology and high energy assumption, touches your heart and brings your simple shock with its sincerity. I like the traumatic literature and the traumatic paintings after Culture Revolution. I like sorrow, powerful things as sword.

Design field is a high-end field with few people, not a field where everyone can enter easily at present day.

作为设计师，应引领市场，而不应被市场所同化，超然毅力和信念的坚守，殉道的精神，这一切都是必不可少的。

室内设计绝对不是装饰设计，它是一个设计师文化底蕴支撑下的空间体验，也可以说是面对空间所产生的一种情愫表达。

我喜欢自在生成的东西，它是抛开了高技、高能耗所产生的一种由感而发的情绪的真诚，即这种空间是带着诚意来感染你，给你的是一种简单的震撼。我喜欢伤痕文学，文革后涌现出来的伤痕画派，喜欢悲怆，喜欢刀一般有力量的东西。

设计师行业是一个高端的少众的行业，而不是现在随意就可以进入的一个领域。

Tian Xi Oriental Club

Design Agency: Shenzhen Dayu Build Space Design Agency
Location: Daya Bay District, Huizhou, Guangdong
Client: Huizhou Jinrunlong Real Estate Development Co., Ltd
Area: 1,500m²
Main Materials: oak, cement self-leveling, proportioning cement

天喜东方会所

设计公司：深圳大羽营造空间设计机构
项目地点：惠州大亚湾区、广东
客　　户：惠州市金润隆房地产开发有限公司
项目面积：1500m²
主要材料：橡木、水泥自流平、配比水泥

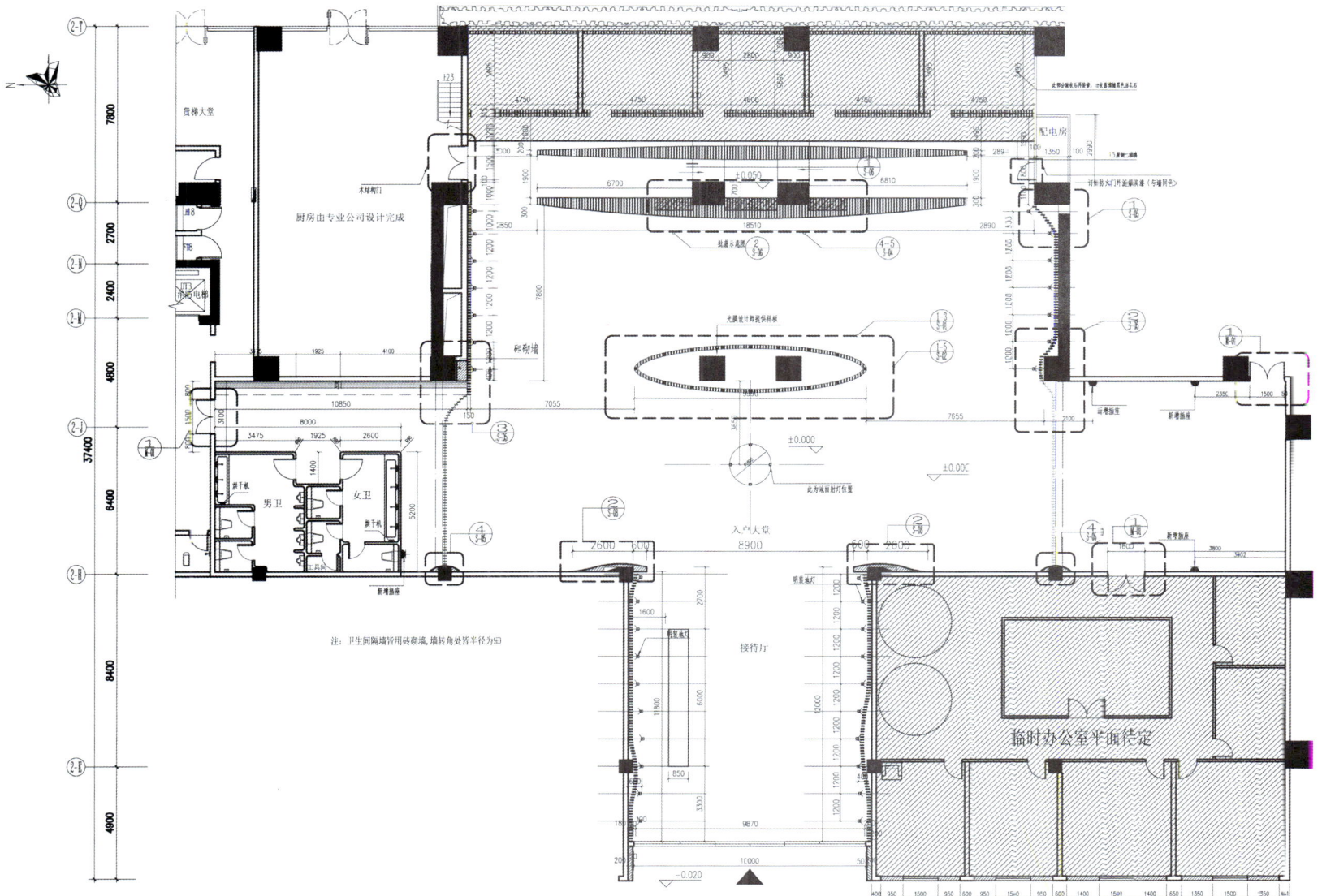

大堂平面图

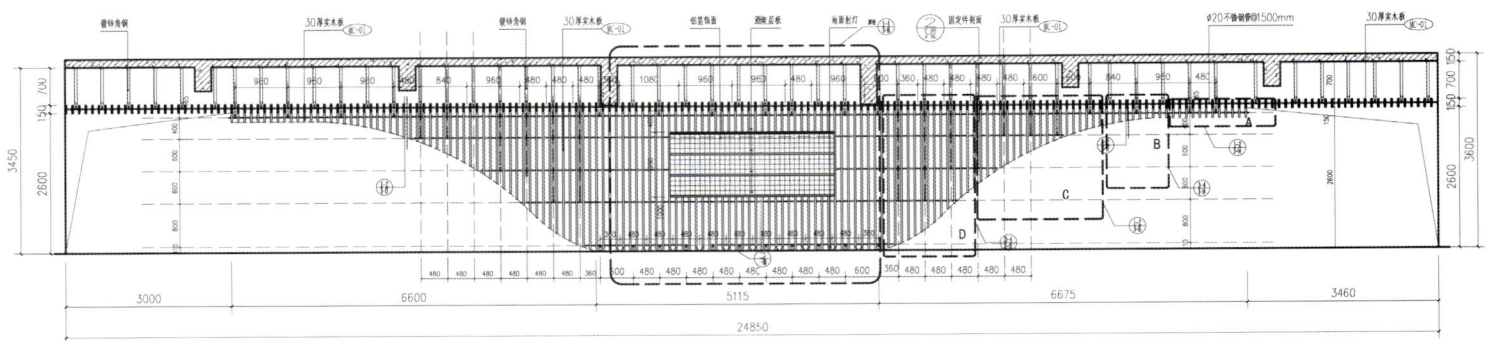

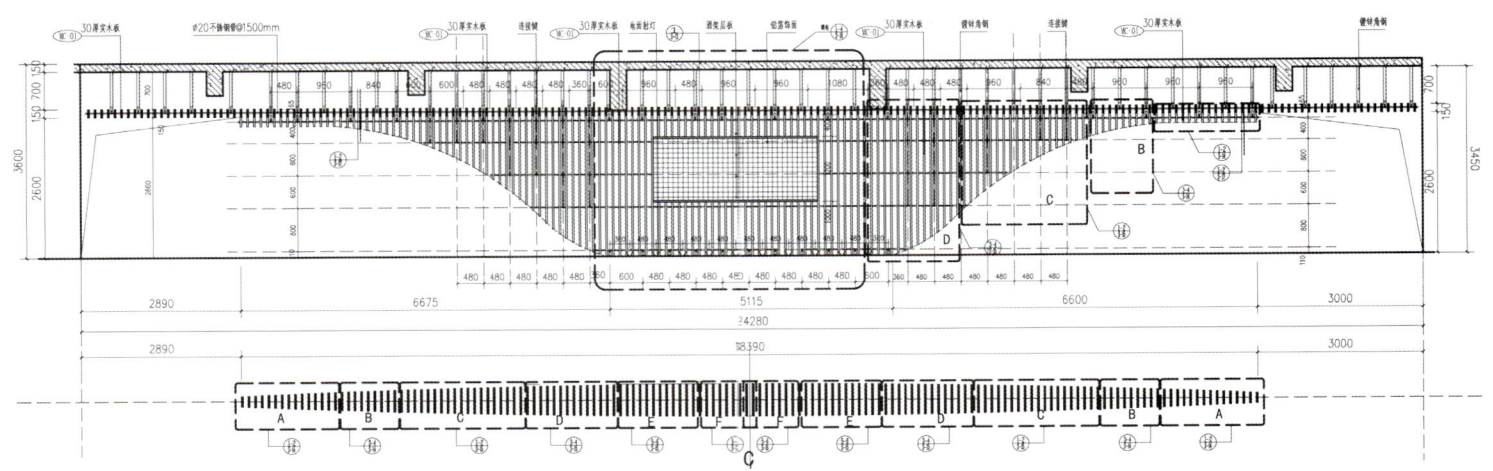

大堂立面图

The idea is from the regional culture relationship of space itself, and as a marketing strategy of the client, in other words, it's oriental thought culture positioning. The two-way culture namely region relation, determines the club hall's space atmosphere: firstly, it's derivative based on the eastern philosophy culture; secondly it explains its regional characteristic in a modern way, namely fully embodying the mountain resources; thirdly it's the original sincerity to respect space, and creates a sincerity space experience.

The shape of whole space originates from an organic form which is wrapped by a double epidermis: the hole rock trace corroded by sea. Bottom is called grass layer which are swing molding metope by using light color cement template, and it means dense vegetation's geographical context relations at the same time, it echoes cultural connotation of eastern philosophy. Top layer uses an organic dimension of rhythmical wood grids to cover the whole space. Hazy, calm and rhythm, this is space image which we want to express.

From the technical construction method, the designers explore to bring the traditional construction technology into interior space which abandons the application of adhesive. The whole construction will be connected with structure and nodes, which could be considered as an interior spacial exploration of tradition regressing. The wall of ground floor uses both grey and white cement at 1:2 ratio. Mixing adhesive, the designers directly processed the 20mm metope with four grass-blade patterns. For the surface layer, the designers handle it from the original smoothly, so as to restore a real, nature feeling of material. Meanwhile, the wooden structure of curve is attached to the surface, making the pattern indistinctly shown. Referencing a regular curve grating is not only a requirement of culture, but remedying the tunnel and structure space regrets of civil engineering.

All in all, from material to technical method, then to shape we fully respect

the truth of material, sincerity of structure, prototype of original construction space, combining culture and creativity, illustrating a brand new modern design exploration of oriented Chinese culture, and this is principle and pursue of human nature.

本案的构思，源于空间本身的地域文化关系和作为甲方的营销宣传的一个策略，即东方思想文化的定位，这种双向的文化即地域的关系，决定了这个会所大堂的空间氛围：一是东方哲学文化基础之上的衍生品；二是现代的诠释其地域特征，即山海资源的充分体现；三是尊重空间的原始诚意性，创造有诚意的空间感受。

整个空间的塑造起源于一个被一个双层表皮所包裹的有机的形态：海水腐蚀礁石后所留下的洞岩痕迹。底层叫做草叶层，是用浅色水泥模板批荡成型墙面，是隐喻周围茂密的草木地理文脉关系，同时与东方哲学文化内涵所呼应。上面一层是用了一个比较有韵律感的木质格栅形态的有机体量覆盖整个空间，"朦胧、宁静、律动"，这是设计师所表达的空间意象。

在工法技术上，设计师探索的是把传统的建造技术引入室内空间当中，整个空间摒弃粘合剂的应用，全部以结构、节点连接所有形体，也可以说是回归传统与一种室内空间探索，包括底层墙壁都用了灰水泥和白水泥，1：2的配比，掺入粘合剂，用四块草叶模板图案，直接对墙面进行了20mm批荡处理，表面处理水泥自流平腊面表层，以还原一种材料真实自然性，形式自然的效果，同时曲面的木质结构附在表面，使图案半隐半现，朦朦胧胧。之所以引用一个律动感的曲面墙体格栅，除了文化底蕴的要求，同时很好地处理了原有土建管道和结构造成的空间遗憾。

总之，从材料到工法再到形体，设计师都尊重材料的真实性，工法与结构的诚意性、原始建筑空间的深度还原，文化的诚意遵守来创造空间的独特气质，全新阐释中国文化背景下的现代设计探索，这是自我本真的一个遵守和追求。

香港方黄建筑师事务所

方 峻

A Hong Kong designer, born in 1970 in Guangzhou, moved to Hong Kong at 15 years old. He successively pursued advanced studies of architecture design and management Bachelor's and Master's course at Hong Kong Polytechnic University, Huaqiao University, Tsinghua University and Italian Milan Polytechnic. He found Hong Kong Fong Wong Architects & Associates in 1997.

Overseas Chinese Town real estate (park entrance VIP reception hall and Wedding Palace), Zhonghai real estate (Longwan Peninsula south club), Zhujiang real estate(sample house), Longhu real estate(the public part), Huarun real estate(the public part of Xiangshuwan) and so on.

　　方峻（TSUN FONG），中国香港设计师，1970年出生于广州，15岁移居香港，先后在香港理工大学、华侨大学、清华大学、意大利米兰理工学院深造，完成建筑设计/管理学士及研究生课程，1997年创办香港方黄建筑师事务所。
　　主要业绩有华侨城地产·公园入口贵宾接待厅及婚礼宫、中海地产·龙湾半岛南会所、珠江地产·样板房、龙湖地产·公共部分、华润地产·橡树湾公共部分等。

Zhonghai · Jincheng Sales Club

Design Agency: Hong Kong Fong Wong Architects & Associates
Location: Ghengdu
Area: 700m²
Photography: Jiang Guozeng

中海 · 锦城销售会所

设计公司：香港方黄建筑师事务所
项目地点：成都
项目面积：700m²
摄影师：江国增

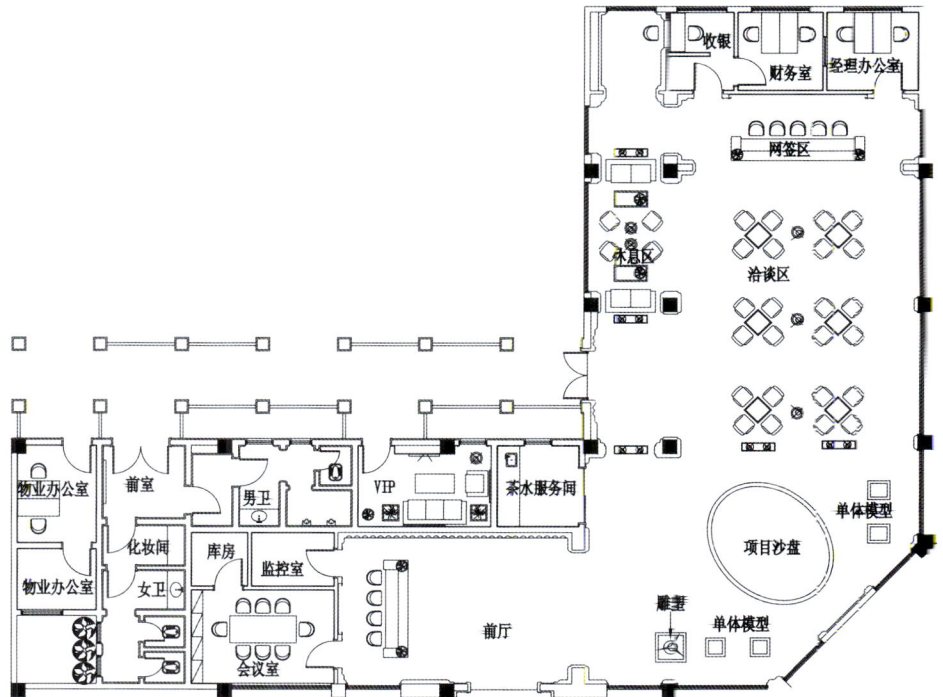

平面图

Because this project is located in the first post of the southern silk road—Cuqiao, the designers traced back to the history of Cuqiao. This area was rich in silk , and trading thriving, once called 'Cocoon bridge', and its brilliant history is originated form the ellipse spheroid—'cocoon'. Through analyzing and refining the form and characteristics of 'cocoon', the designers designed the unique pattern which is used in this project of interior design, bringing beautiful prospect of chrysalis into butterfly, and singing the song of modern city.

Analysis of design: the designers blend the theme pattern into the place filled with the new art adornment style, make function and aesthetics in one, and use modern visual elements and stone full of sense of quality, tile, wallpaper, metallic paint and other materials as people who play the role of storytelling in the club room to tell the story about local historical development and urban evolution. Through the theme of 'cocoon' shadow, the idea of 'What is club, where is club from, where club wants to go' is represented.

　　由于该项目是在南方丝绸之路的第一个驿站——簇桥，于是，设计师追溯簇桥的历史渊源——该地在历史上盛产蚕丝，且蚕丝交易兴旺，曾一度被称为"茧桥"，其辉煌的历史就是源自于千丝万缕叠积而成的椭圆形球体——"茧"。因此，设计师通过对"蚕茧"形态与特征的分析与提炼，形成独特的主题纹样，运用于该项目的室内设计之中，希望带给这里化蛹成蝶的美丽意境，在这里唱响当代现代化都市生活的序曲。

　　设计构成分析：把主题纹样图案融入新艺术装饰主义风格的空间之中，让空间集功能与审美于一体，让现代视觉元素和富有品质感的石材、地砖、墙纸、金属漆等材料在会所空间中扮演说故事的人的角色，讲述当地历史发展与城市演变的故事。通过"茧"影这一主题，轻盈优雅地道出："会所是什么？会所从哪里来？会所要到哪里去？"

昊泽空间设计有限公司

韩 松

Han Song is graduated from Hubei Academy of Fine Arts, Department of Environmental Art and Interior Design in 1997 and he is the general manager and design director of Shenzhen Horizon Space Design Co.,Ltd. This company started partnership with Vanke in 2007 and has continued sample house design strategic cooperative partnership in 2009 and 2010 since 2008. They are mainly responsible for decoration design of sample house for Vanke Shenzhen area houses. With many years' cooperation with Vanke, they accumulate a wealth of experience in designing sample house in the event of fine decoration.

They have a group of young and high-quality professional technicians and supporting professionals as backing force and they are a professional design community with a strong sense of team spirit and cooperation with foreign design agency. They have cooperation with BUROII company of Belgium and finished the design of Guangzhou Baiyun International Convention Centre in 2007, cooperation with HBA company of the USA and finished the design of both Beijing Ritz-Carlton Hotel and Beijing Wanhao Hotel, cooperation with Line And Space company of the USA and finished the design of Vanke Tianqinwan Villa.

1997毕业于湖北美术学院环境艺术及室内设计系；现任深圳市昊泽空间设计有限公司总经理及设计总监。公司于2007年开始和万科地产合作，并于2008年起，分别在2009、2010年与万科延续样板房设计战略合作伙伴关系，主要负责万科深圳区域楼盘样板房的精装修设计。多年与万科的设计合作让公司在楼盘样板房的精装修设计上积累了丰富的经验。

公司拥有一批年轻的高素质的专业技术人才和配套专业人员做后盾，是有强烈团队意识和合作精神的专业设计群体。公司与境外的设计公司开展合作，2007年与比利时BUROII公司合作，完成广州白云国际会议中心深化设计；与美国HBA公司合作，完成北京丽兹卡尔顿酒店深化设计；与美国HBA公司合作，完成北京万豪酒店深化设计；2009年与美国Line And Space公司合作，完成万科天琴湾别墅的深化设计。

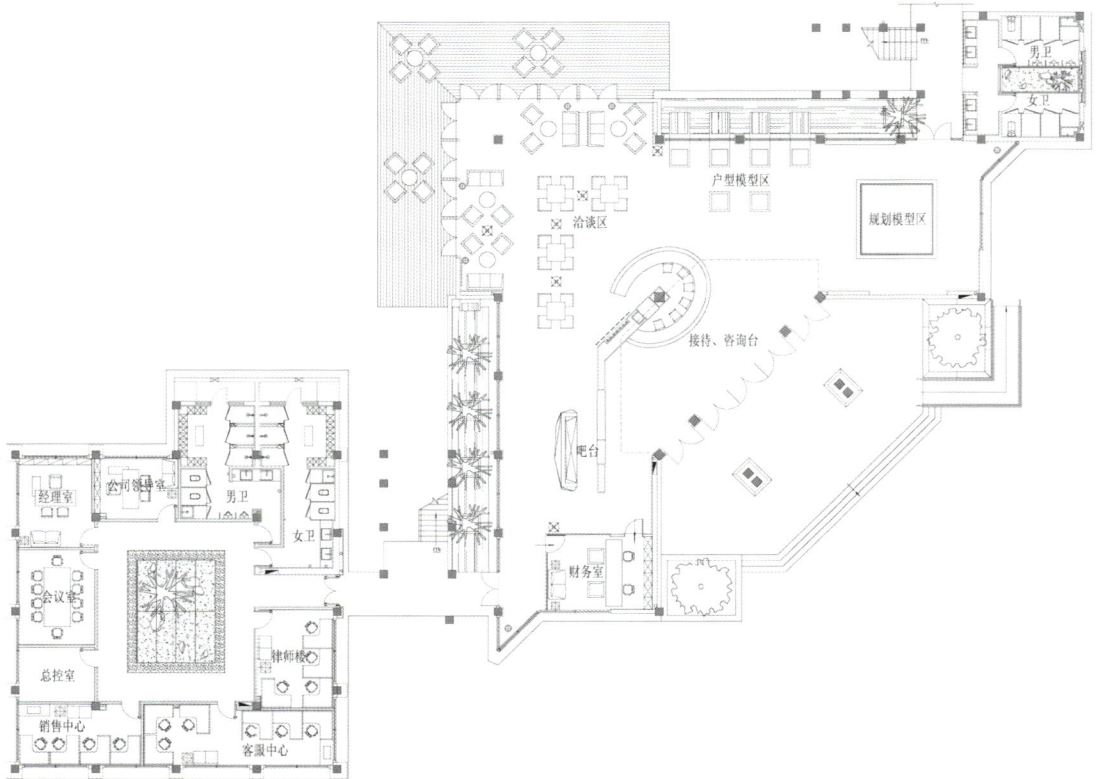

平面图

Futong Real Estate Tianyiwan Sales Office

Design Agency: Horizon Space Design Co.,Ltd
Location: Dongguan
Client: Dongguan Futong Zhuchuang Real Estate
Development Co., Ltd
Area: 620m²
Photography: Xiao Deng
Main Materials: ancient wood grain stone, white sand
cream-colored stone, teak board, black mirror

富通地产天邑湾项目售楼处

设计公司：深圳市昊泽空间设计有限公司
项目地点：东莞
客　　户：东莞市富通铸创房地产开发有限公司
项目面积：620m²
摄 影 师：小邓
主要材料：古木纹石材、白砂米黄石材、柚木板，黑镜

This project is located in Dongguan City, and the engineering cost is 3000yuan/m². 本项目位于东莞，工程造价3000元/平方米。

1.The architecture emphasizes natural growing structure, and it makes full use of existing advantage of space.

2.The integral style pursues integration of tradition, nature and modern aesthetic standard, so as to showcase the whole architecture.

3.The permeability of space and the extension of natural landscape increase the space visual advantage and enhance the psychological value of space.

1. 建筑强调自然生长结构，充分利用现有建筑空间的优势。

2. 整体风格追求传统、自然和现代审美的融合，让建筑完美地展现出来。

3. 空间的通透性和对外部自然景观的引伸，增加空间视觉优势，提升空间心理价值。

Headquarters Office Signing Centre of Meilin Vanke

Design Agency: Horizon Space Design Co.,Ltd
Location: Shenzhen
Client: Shenzhen Vanke Real Estate Co., Ltd
Area: 262m²
Photography: Han Song
Main Materials: light maple science and technology wood, carpet, frosted glass

万科梅林总部办公室签约中心

设计公司：深圳市昊泽空间设计有限公司
项目地点：深圳
客　　户：深圳万科房地产有限公司
项目面积：262m²
摄 影 师：韩松
主要材料：浅枫科技木、方块地毯、磨砂玻璃

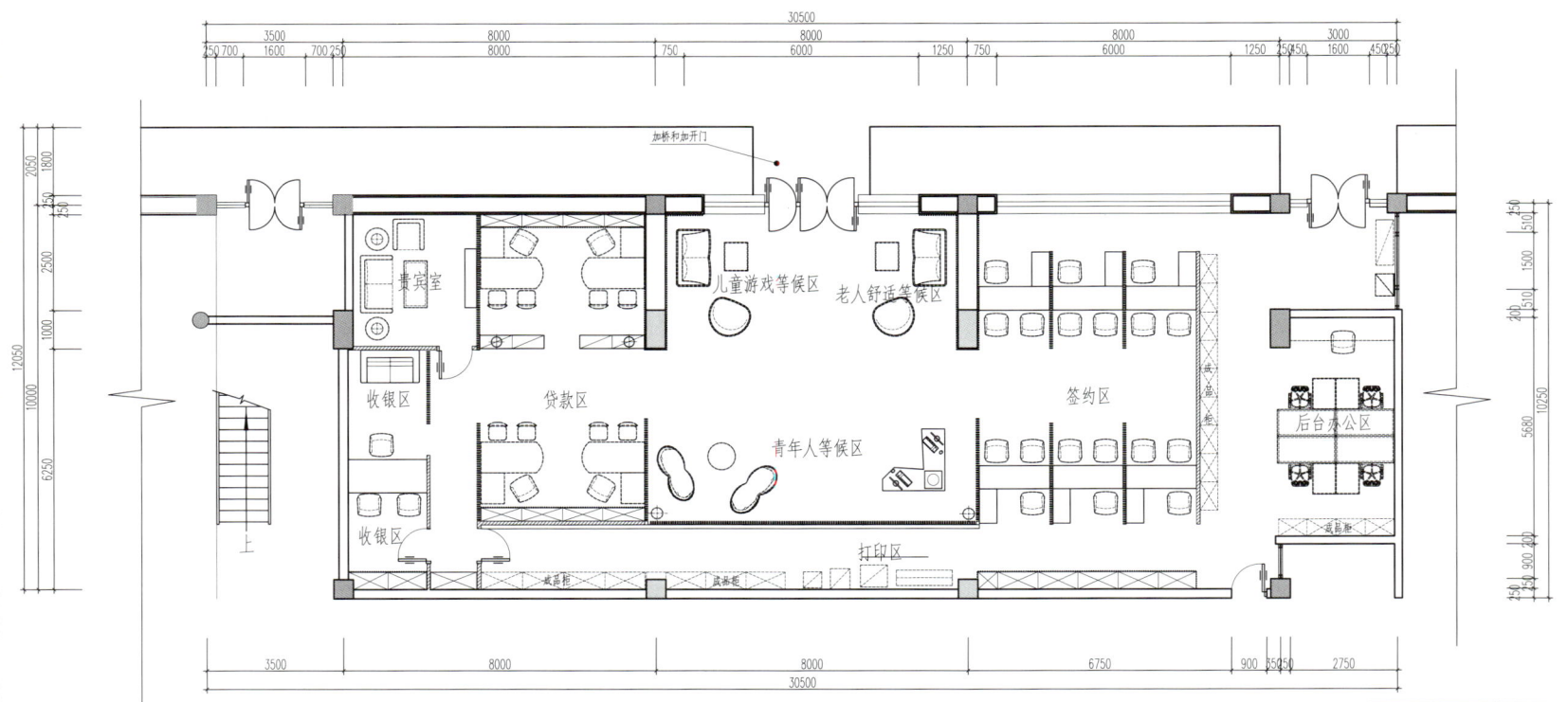

平面图

图例	名称
	原建筑墙体
	轻钢龙骨隔墙
	现砌筑墙体

Headquarters Office Signing Centre of Meilin Vanke is regarded as the best signing centre of Vanke. Although purchasing a house is very uncomfortable, but at least buyers can feel free and have a short period of happiness when they sign the contract.

　　万科梅林总部办公室签约中心，被评为万科最好的签约中心。买房子越来越痛苦了，至少在签约的时候，心态是祥和的，让购房者拥有短暂的幸福感吧。

Zhongshan Vanke Langrun Garden Sales Office

Design Agency: Horizon Space Design Co.,Ltd
Location: Zhongshan
Client: Dongguan Xin Wan Real Estate Development
Co., Ltd
Area: 370m²
Photography: Chen Zhong
Main Materials: cable color rosewood, antique profile
cream-colored stone, edged mosaic stone, solid wood
flooring, hand paint wallpaper

万科中山朗润园售楼处

设计公司：深圳市昊泽空间设计有限公司
项目地点：中山
客　　户：东莞市新万房地产开发有限公司
项目面积：370m²
摄 影 师：陈中
主要材料：花梨木索色、米黄石材仿古面、石材马赛
克镶边、实木地板拼铺、手绘墙纸

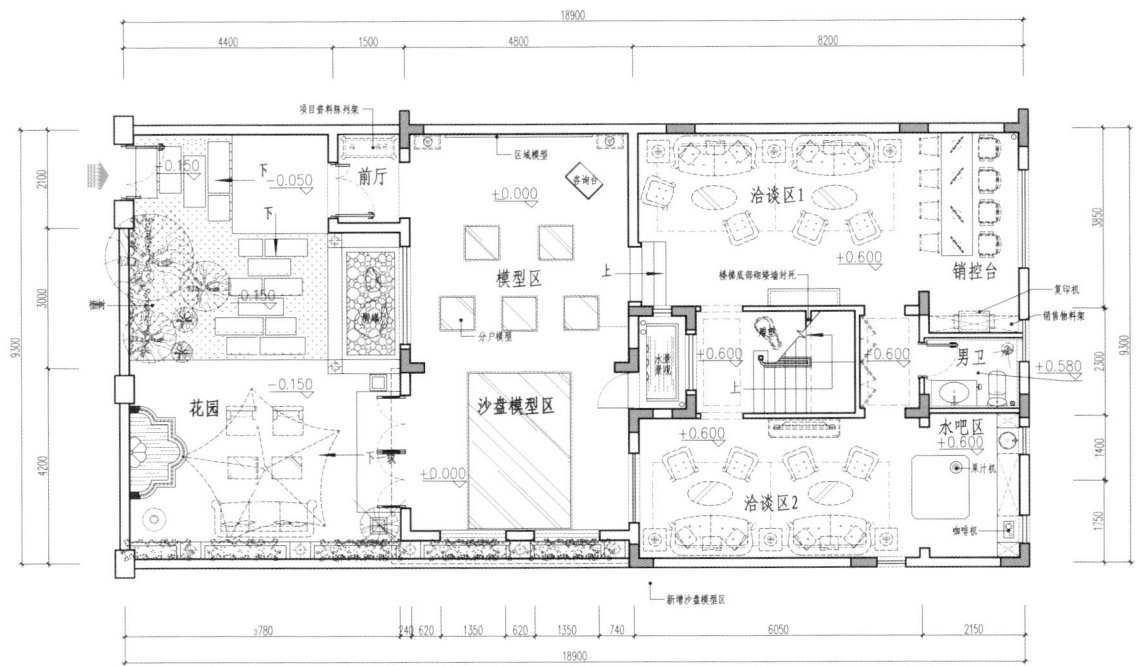

1F平面图

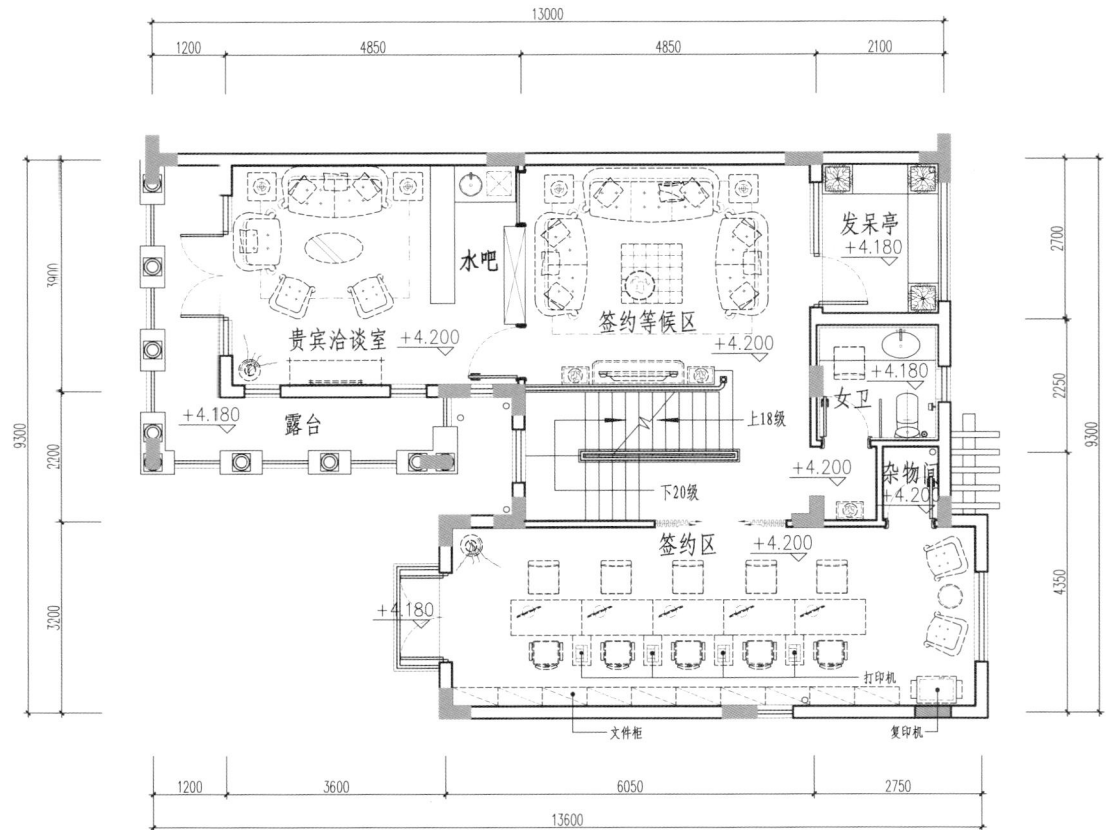

2F平面图

Since the houses are villas, the sales office emphasizes specificity and honor sense of customer.

因本楼盘为别墅群，售楼处强调专属性和客户的尊贵感。

本书参编人员（排名不分先后）：

邹志雄	戴　勇	李益中	黄士华	冯　羽
邱春瑞	刘　鹏	赵牧桓	王春添	邱宜平
黄怀德	方　峻	赵力行	琚　宾	洪德成
刘晓都	孟　岩	王　辉	刘卫军	韩　松
张质生	黄鹏霖	林宪政	李　泷	

卢积灿	王　创	王　宇	张喜娜	王晓冬
罗　斌	肖湘桂	陈桂娇	洪　辉	杨思雁
陈　龙	项国媚	黄　丹	王　佳	丁小娟
李　莹	李　忍	李　静	鲍　威	李思瑜
彭　莹	罗　方	胥群燕	李文珠	刘莉莉
洪　霖	江　勇	马志远	王佳丽	雷　云
杨　傲	沈　婷	胡晓意	肖珍华	张　蕾
杨创创	冯　韬	李继旺	周　晶	陈思锦
李春华	李姣姣	申　琦	肖广俭	曹晓云
梁艺芳	黎　芳	林宗明	梁春苗	李　强
王　伟				